The San Juan Triangle of Colorado

Mountains of Minerals

Contributors

David W. Bunk, Robert B. Cook, Don Earnest, Robert W. Eveleth
Daniel Kile, Virgil W. Lueth, Gary McWilliams
Barbara L. Muntyan, Jack A. Murphy, Horace B. Patton
Tom Rosemeyer, Duane A. Smith, Gloria A. Staebler
Kaitlin Stoufer, Robert Stoufer, John Trujillo, Karen A. Vendl
Mark A. Vendl, Terry C. Wallace and Allan Young

Editors

Daniel Kile and Gloria A. Staebler
with David W. Bunk, Tom Hughes and Suzanne Liebetrau

Illustrators

Michael Bainbridge, Dan Behnke, William W. Besse, Jack Brauer, G.V. Buck
Joe Budd, Paul Carrara, Chip Clark, Jennifer Cochran, Joseph Collier
Bill Dameron, Frederick Endlich, Charles Goodwin, William Henry Jackson
Daniel Kile, Jesse La Plante, Barbara L. Muntyan, Jack A. Murphy, John Rahart
Tom Rosemeyer, Jeff Scovil, J. Walter Socha, John Smolski, Gloria A. Staebler
Robert Stoufer, Gene Tribbey, Ogden Tweto, Mark A. Vendl and Allan Young

Lithographie, LLC • Denver

No. 15: The San Juan Triangle of Colorado
Mountains of Minerals

ISBN 978-0-9836323-0-6

Publisher
Lithographie, LLC
Denver, Colorado, USA

PO Box 11613 • Denver, CO 80211 • USA
Phone: 1.303.495.5521 • Fax: 1.303.482.1238
www.lithographie.org

Printed in the United States of America

Photo Credits

Cover: Rhodochrosite, 6 cm wide, Sunnyside Mine, San Juan County, Colorado; Steve Smale collection, Jeff Scovil photo

Title Page
Background: "Tram from Gold King to the Mill" in San Juan County, Colorado; San Juan County Historical Society Archive collection, circa 1880s
Inset: Faceted sphalerite, 16.87 carats, Camp Bird Mine, Ouray County, Colorado; Dave Bunk collection, Jeff Scovil photo

Background: A late afternoon view of Red Mountain #2 as seen from the summit of Red Mountain #3
Photo Jack Brauer/WideRange Photo LLC

Facing Page Insets
Top: Gold, 4.6 cm tall; Smuggler Union Mine, Telluride, San Miguel County, Colorado; Colorado School of Mines collection, Jeff Scovil photo
Upper Middle: Quartz and chalcopyrite, 6.5 cm tall, Black Bear vein, Dave Roter collection, Jesse La Plante photo
Lower Middle: Rhodochrosite, 3 cm tall; Mountain Monarch, Ouray County, Colorado; Tom Hughes collection, Jesse La Plante photo
Bottom: Hübnerite, 7.5 cm tall; Adams Mine, Silverton, San Juan County, Colorado; Dave Bunk collection, Jeff Scovil photo

Contents

4 *The San Juan Triangle of Colorado*

Ho for the San Juans!

by Duane A. Smith, Mark A. Vendl, and Karen A. Vendl

Located in southwestern Colorado and covering over 31,000 square kilometers, the San Juans are some of the highest and most rugged mountains in the state. Hundreds of peaks rise above 13,000 feet, including about a dozen that exceed 14,000 feet in elevation. For the past 250 years, men have sought to extract gold and silver from these mountains. However, it was not until the last 25 years of the 1800s that the San Juans began revealing the extent of their mineral wealth. In addition to gold and silver, the region produced economic quantities of lead, zinc, tungsten, and copper. In 1896 world-renowned mining engineer Thomas "T.A." Rickard called the San Juans one of Colorado's four great mining regions along with Leadville, Cripple Creek, and Central City (Smith, 1996).

The Spanish, during the eighteenth century, were the first to mine in the San Juans. One of the earliest recorded expeditions to the area was led by Juan Maria de Rivera in 1765. By that time many of the Spanish place names were in use. A 1776 expedition led by Atanasio Domínguez and Silvestre Vélez de Escalante skirted the southern edge of the San Juans, and Father Escalante wrote in his journal,

> *They say there are veins and outcroppings of metal ... The opinion formed previously by some persons from accounts of various Indians and of some citizens of this kingdom that there were silver mines caused the mountain to be called Sierra de la Plata* [Silver Mountains].

Spanish miners had spent a season or two in the San Juans but avoided recording their efforts because they were trespassing on the King's resources and wanted to avoid paying one fifth of their profits—the "royal fifth"—to the Spanish Crown. Early Anglo miners found ample evidence of the Spanish workings and tools throughout the San Juans (Smith, 2000).

Fur trappers working the San Juans in the early 1800s were aware of the area's mineral potential. Well-known mountain man and guide, Kit Carson "strongly insisted" that these southern ranges were "prolific in mines of gold, silver, and other precious stones." By the 1859 Pikes Peak gold rush, legends and tales of the San Juans were circulating among the newly arrived miners. Initially, all of the activity was concentrated along the eastern slope of the Rockies, but as areas such as Central City and Idaho Springs were "staked out," prospectors fanned out to the west.

In August 1860, Charles Baker led a small party into the San Juans. That fall in the *Rocky Mountain News* (October 12,

---

Above: August sunrise on middle Blue Lake, Ouray County
Photo Jack Brauer/WideRange Photo LLC, 2010

Facing page: Pen and ink and watercolor map of the Domínguez-Escalante 1776 expedition through Colorado, Utah, Arizona and New Mexico

Left: "Pandora Miners, Telluride, Colo" circa 1880s "Byers photo" Telluride Historical Museum collection

Facing page: The large, partially collapsed Frisco (Bagley) mill building just outside of Animas Forks Jennifer Cochran photo, 2008

1860), one of the members of his party described the San Juans as "the highest, roughest, broadest and most abrupt of all ranges." He further observed that in this range, "the metalliferous development of this region, if not of the North American continent, reaches its culminating point." That was all it took—the rush to the San Juans was on.

The summer of 1861 drew as many as 1,000 men to Baker's Park (now the location of Silverton, Howardsville, and Eureka). Unfortunately, the rush failed because of the lack of placer gold, unfriendly Ute Indians, isolation, and the short prospecting season afforded by the high-elevation climate. It could snow any time of year. As an old miner declared, the climate produced three months of mighty late fall, and nine months "of damn hard winter" (Smith, 1996). By the fall of 1861, the prospectors had left. The Civil War consumed the country's interest, and the San Juans were quiet for the remainder of the decade.

In 1870 prospectors returned to Baker's Park, where they staked claims but initially worked them only in the summer. They were no longer searching simply for placer gold; lodes were discovered, mines were opened, and by 1874, small mining camps had begun to appear. The towns of Silverton, Howardsville, and Eureka took hold, providing services to the miners.

The Little Giant, developed in 1870 and located at the head of Arrastra Gulch near Silverton, was the first lode mine in the San Juans. Its gold assays ran from $1,000 to $4,000 a ton. For the next two years, it grabbed headlines and was sold in 1873 to Chicago investors for half a million dollars. Never achieving its predicted success, the Little Giant was the region's first great failure.

For more than 500 years, the San Juan Mountains had been the summer home of the Ute Indians, but the miners threatened the Ute's way of life. Initially, the government tried to uphold the agreement with the Utes by keeping miners off their land, but the lure of gold made that impossible. In September 1873, the Brunot Agreement was signed, ceding 3.5 million acres of Ute land—the heart of the San Juan mining country—to the United States. In return, the Indians were to receive $25,000 annually and retained the right to peacefully hunt in the San Juans (Smith, 1996). With the signing of the Brunot Agreement, it was legal for miners to be there, and they poured into Baker's Park, staking more than 2,000 claims by the end of 1873. Many of the area's famous properties were discovered at that time, the Sunnyside, Shenandoah, Silver Lake, and Gold King (Smith, 2004) among them.

Three ingredients are essential to building a major mining region: profitable mineralization, capital, and economical and reliable transportation. For the miners the 1870s were a decade of waiting: they knew that the mineralization was there, but they needed investors to provide the funds, mills, and smelters, and most of all, they needed efficient transportation. Author Frank Fossett (1879) described the situation:

> *Promising as were the numerous discoveries of the San Juan country in 1873-74-75, they were generally of no immediate benefit to their owners, on account of the distance from an ore market and smelters, wagon roads, and railways. The region labored under peculiar disadvantages. It was made up of almost inaccessible mountain ranges, and at that time, was so remote that capitalists and mill men were not inclined to investigate its mineral wealth. The pioneers who had been making discoveries of rich veins were too poor to build works for extraction of the precious metals, and it cost too much to get ore to the market.*

Slowly, eastern capitalists began to invest in the mines, and by the end of the decade, William Palmer and his Denver

& Rio Grande Railroad had set its sights on the San Juans. Palmer had described the area as having "more rich silver lodes" than had been discovered in any western state or territory. His plan was to build a narrow gauge to run from the south up Animas Canyon to Silverton. In September 1880, the Denver & Rio Grande settled on the new town of Durango, at the southern edge of the San Juans, as the location for its facility. Within three months, 2,500 people had crowded into the new town, which eventually became the regional smelting center; the New York and San Juan smelter provided the most advanced technology available (Smith, 1996). In July 1882 to great fanfare, the first train steamed into Silverton. The Denver & Rio Grande continued building railroads in the San Juans, arriving in Ouray in 1887 and Lake City in 1889.

However, the Denver & Rio Grande was not the only game in town. Businessman Otto Mears, who had built many of the toll roads in the San Juans, decided to get into the railroad-building business to provide service to the mines near Silverton. From 1888 to 1889, Mears built the Silverton Railroad to connect the Red Mountain area to Silverton. In 1896 he began constructing the Silverton Northern Railroad, which reached Animas Forks in 1904. In 1915 Mears purchased the Silverton, Gladstone and Northerly Railroad, which had reached Gladstone in 1899. All three of these narrow gauge lines connected to the Denver & Rio Grande Railroad in Silverton, thus allowing ore to be transported to Durango for smelting. In 1890 and 1891, Mears also built the 172-mile Rio Grande Southern, which connected Durango to Telluride and Ridgway. With the railroads dramatically reducing the cost of shipping, lower-grade ores could be processed economically. Mears had intended to extend his Silverton Railroad past the Red Mountain area to Ouray. However due to the steep grades in the Uncompahgre Canyon, the extension was never constructed.

The San Juans were settled in a series of small rushes to Lake City, Ouray, Telluride, and anywhere else that held the faintest trace of mineralization. Animas Forks, Howardsville, Rose's Cabin, Ames, Mineral Point, and other towns all had experienced flurries of excitement. Prospectors and miners scoured every recess of these mountains looking for their fortune. Most never found it.

In the 1880s the Red Mountain District (see page 16) captured the spotlight in the San Juans. Located between Silverton and Ouray, Red Mountain was initially prospected in the late 1870s. However it was not until the 1882 discovery of the Guston and Yankee Girl Mines that Red Mountain gained national attention. Ore in both mines proved rich right from the "grass roots," so rich that it was not sorted, just sacked and shipped out. From 1884 into the 1990s, the mines of Red Mountain produced steadily, especially after the Silverton Railroad reached the district.

The crash of 1893 and the subsequent depression caused silver values to drop, which forced many of the silver mines to close. Fortunately the San Juans still had gold mines, particularly within the triangle defined by Ouray, Silverton, and Telluride. This area includes such famous mines as the Tomboy, Smuggler-Union, Camp Bird (page 30), and Liberty Bell. These mines continued to produce well into the twentieth century. The last major mine, the famous Sunnyside (page 68), did not close until 1991.

In the 1890s one of the greatest mines in United States' history was discovered and developed southwest of Ouray: the Camp Bird. For its first decade, the Camp Bird not only maintained high gold production, but also used the best current mining methods and equipment, attracted visitors from around the world, and made its original owner, Thomas Walsh, the region's first millionaire.

Telluride languished until the arrival of the Rio Grande

The Blue Sky People

by Kaitlin Stoufer

Left: Circa 1899 photomechcanical print entitled, "Utes—Chief Sevara and family"
Photo by Nast of Denver
Library of Congress Prints and Photographs Division

Nomadic independent bands of Ute Indians roamed western Colorado and eastern Utah as well as parts of New Mexico and Wyoming for centuries before their first contact with westerners—the Spanish—in the late sixteenth and early seventeenth centuries. Pushed into the mountains from the Four Corners area by the Plains Indians, the small bands of Utes hunted in the San Juans in the summer and spent their winters with larger groups in the lower, warmer foothills. They hunted deer, elk, buffalo, and (mainly) jackrabbit and gathered vegetation. They knew their revered San Juans as "the Shining Mountains."

Early interactions with the Spanish brought some fighting but soon settled into a solid, trade-based relationship. Historians agree that the Utes were among the first Indians to be introduced to horses and to the many advantages they brought. With the aid of horses, the Utes could travel farther and in larger groups. They thus began to spend less time in the mountains. But horses also brought the Utes a taste for livestock instead of game and a means of stealing food instead of hunting for it, bringing obvious harm to their relationship with their white neighbors.

The Utes created landmarks to help the Spanish navigate Ute territory; their early trails are the basis for many of Colorado's modern roads. The expanding Spanish presence in the area on one hand left the Indians uneasy, but their discomfort was offset by the benefits of their expanding trade economy. These were prosperous times for the Utes. With less time spent hunting, they were free to develop culturally, building on traditions of music, dance, and religion; they even enjoyed horse racing! Although they preferred to stay in small groups, Ute families did form informal alliances with one another for defense. By the late eighteenth century, seven loosely organized Ute nations inhabited Colorado, Utah, and New Mexico. Some of their chiefs tried unsuccessfully to bring the nations together; still, the Ute nations lived peacefully with one another.

The end of the US-Mexican War in 1848 brought an influx of settlers to Colorado and predictable changes to the relationship between the whites and the Indians. Treaties restricted Ute movement and territory, and the permanent settlements, fences, and farming customs of the white men were foreign, even offensive, to the migratory, nature-worshipping Utes, known to the whites as "Blue Sky" people.

The 1858 discovery of gold near present-day Denver caused a rush to the area that further marginalized the Utes. In 1863 with tensions mounting, Ute and Colorado leaders met to discuss the possibility of moving the tribes. The Utes refused to relocate, but they agreed to stay within specified boundaries. In return, they were promised livestock and supplies. Of course, the restrictions on their activities and the reduction of territory limited their hunting opportunities, rendering the Utes poorer and more reliant on government supplies—which by most accounts never materialized.

Over the next few years, the Utes were moved onto a reservation that stretched from today's Gunnison to an area west of the Utah line. But in 1870, the Baker party found gold near present-day Silverton—undisputed Ute territory.

Businessman Otto Mears and Ute Chief Ouray helped broker the 1873 Treaty of Brunot, which required the Utes to once again cede their land (including the San Juan Mountains) and relocate. The Indians retained the right to hunt in the San Juans as long as they did so without confronting the settlers. The government agreed to pay the Indians an annual sum of $25,000, but as seems to have been their habit, Congress did not begin distributing the promised funds until an Anglo-Indian war appeared inevitable.

Some Utes believed that Chief Ouray was receiving but not sharing the stipend. Tensions, both internal and external, were once again on the rise. The territorial government lobbied Congress to pay the Utes; war was averted, and in 1881, the Utes were moved to a permanent reservation in northeastern Utah.

The new land provided very little game, and the Utes were allowed to return to the San Juans each fall for a few years until, in 1887, they were permanently barred from hunting in the Shining Mountains.

Ames Power Station
In 1890-91, L.L. Nunn built the world's first single-phase alternating-current power transmission system at Ames, Colorado in order to power the stamp mill at the Gold King Mine, 2.6 miles away. The plant consisted of a Pelton water wheel connected to a 3000 volt, 133 cycle Westinghouse AC generator. The current was then transmitted to the stamp mill. The power station was upgraded in 1895 (building to the right of center) and again in 1905 (center building). Library of Congress Prints and Photographs Division

Southern on November 23, 1890. The Tomboy, Smuggler-Union, and Liberty Bell mines, along with their lesser neighbors, brought Telluride into its heyday in the 1890s. Silverton enjoyed its long-awaited boom when the Silver Lake and Gold King mines came into their own in the 1890s.

Beginning in the 1870s, the San Juans were opened and developed, the towns supporting them were settled, and a solid infrastructure was developed. Production figures for these years gives an idea of the economic impact of these mines, but it was more than equaled by the social impact. San Miguel County produced over $70 million in gold, silver, lead, zinc, and copper; San Juan County produced $53 million; and Ouray County, $67 million.

In the San Juans, advances were made in the use of tramways to transport ore and in the use of electricity in mining operations. Tramways, bucket-carrying cables supported by towers, stretched from mine to mill or railroad siding, thus offering unmistakable advantages in year-round operation. With gravity's assistance, tramways overcame topographic obstructions and lowered costs; the weight of the full buckets of ore coming down pulled empty buckets up. Trams could be constructed almost anywhere—over cliffs, up mountains, across streams—and in theory were cheaper to maintain than roads. A snow slide, however, might carry away a couple of towers and mitigate that advantage. Versatile, tramways carried ore out and brought supplies in, and venturesome miners traveled relatively quickly on them. Swaying to and fro in a bucket, numbed by the cold wind, many tram-riders probably wondered if it would have been wiser to walk. Because of the ruggedness of the San Juans, tramways became very popular between 1890 and 1920. In fact, the San Juans have been called the tramway capital of the world, with at least two dozen significant installations (Trennert, 2001).

A few far-sighted San Juaners begrudged the waste of unharnessed water tumbling over falls. Mine owner J.S. Robinson, writing his Leadville (Colorado) friend Henry Moody, considered the water issue, "Could not this water power be converted into electricity?" Indeed it could; the guiding genius who made it happen was Lucien Nunn, an ambitious lawyer-turned-mining man. As manager of the Gold King Mine in Telluride, Nunn lost money because of the expense of freighting fuel. He evaluated several means of reducing costs, including the use of direct electrical current. However, Nunn settled on experimenting with alternating current, which could transmit higher voltage more cheaply and easily than could direct current. Nunn visited Westinghouse Electric Company in Pittsburgh, Pennsylvania and after a series of experiments, began construction of a plant at Ames, Colorado. In the spring of 1891, the generator revved up, and electric current raced to the Gold King Mine and mill, running nonstop for thirty days. This was the first time alternating current had been transmitted over a long distance. The stockholders of the Gold King quickly realized the implications of this innovation, which reduced expenses and put the property into profit. Soon other mines including the Tomboy and Camp Bird recognized the advantages of electricity and starting using alternating current.

The story of the San Juans is not only about famous mines, mining technology, and high-grade ore, but also about the pioneers who settled these rugged mountains. They came by the hundreds, then by the thousands, always searching, forever seeking a dream. Unsung and unknown today, they labored, played, succeeded, and failed.

For those who want to delve into the area's history more deeply, the reader is directed to author Duane Smith's book, *The Song of the Hammer and Drill* (2000).

The mountains are calling me and I must go.
–John Muir (1838–1914), naturalist

Right: The eroded Fish Canyon volcanic tuff was produced by the eruptions of the La Garita Caldera and is exposed in the remote Wheeler Geologic Area, near Creede. The area is named for First Lieutenant George Montague Wheeler, who explored and surveyed this area in 1874 as part of the U.S. government's ambitious plan to map the area west of the 100th meridian.

The hardened ancient flow is shown here at sunrise in September, 2006.
Photo Jack Brauer/WideRange Photo LLC

Facing Page: The rugged terrain of the San Juans is evident from the ridge of Mount Sneffels (14,150 feet), near Ouray
Photo Jack Brauer/WideRange Photo LLC
May 2008

Geology of the San Juan Mountains

by Terry C. Wallace and Tom Rosemeyer

The San Juan Mountains are a spectacular range of towering and rugged peaks that cover about 25,000 square kilometers of southwestern Colorado. Stretching from Creede in the northeast to Durango in the southwest, the San Juans are home to some 14 peaks over 14,000 feet in elevation and literally hundreds of peaks that top 12,000 feet. The topography is extraordinarily steep, and much of the range is above timberline. The imposing landscape was shaped by some of the most violent volcanic eruptions known in geologic history. Between 35 and 26 million years ago, huge volcanic calderas rose and collapsed and erupted tens of thousands of cubic kilometers of rhyolitic and andesitic tuffs. The scarred landscape that remained was full of fractures and faults that would later localize the magmatic fluids that deposited the ore bodies of some of Colorado's richest mining districts: Creede, Lake City, Summitville, Silverton, Ouray, Telluride, and Rico.

Although the volcanic history of the San Juans is most strongly etched on its physiographic features, billions of years of geologic history are exposed in these mountains. The oldest formations found in the range consist of a basement complex of Precambrian metasedimentary and metavolcanic rocks. These rocks, which are approximately 1.8 billion years old, represent some of the earliest continental crust preserved in the continental United States and are the expression of just one period of intense mountain building within the Colorado Rockies. The Precambrian mountains slowly eroded and were eventually covered by a shallow sea in the Paleozoic. For hundreds of millions of years, the oceanic waters precipitated a very thick sequence of limestones interlayered with sandstones, shales, and conglomerates. In the western San Juan Mountains along the drainage of the Animas River, this sequence of sedimentary rocks is exposed as the towering cliffs of the Hermosa formation and the Leadville Limestone cliffs surrounding Ouray.

There were several episodes of uplift and erosion forming the San Juans throughout the Mesozoic era, but the modern face of the San Juan Mountains began to emerge during an extraordinary episode of volcanism that began about 35 million years ago. More than 30 centers of volcanism formed throughout southern Colorado and northern New Mexico in what is known as the Mid-Tertiary Ignimbrite Flare-Up. These volcanoes probably looked like stratavolcanos, which form above subduction zones (eg., Mount Rainier and Mount Fuji), but they produced far more voluminous eruptions. Initially the eruptions produced andesites and explosive ash falls, but starting about 30 million years ago, huge sheets of pyroclastic flows were extruded. The pyroclastic flows are

Geologic Map of the State of Colorado
scale 1:500,000
Ouray, San Juan, and San Miguel counties are in the southwestern portion of the state
Ogden Tweto map, 1979

welded tuffs known as *ignimbrites*. They are unparalleled in size; within the San Juans there are at least 22 flows that are larger than 100 cubic kilometers. The only way to explain them is to assume nearly continuous eruptions for dozens of years. The volcanic centers ultimately collapsed, forming large calderas. The largest eruption known in the geologic record anywhere in the world occurred in the San Juan Mountains at the La Garita Caldera, north of Creede. La Garita produced the Fish Canyon Tuff 28 million years ago; this tuff is more than 5000 cubic kilometers and is a 25-by-75-kilometer oblong flow of dacitic composition. For scale, the Fish Canyon Tuff could fill Lake Michigan! Equally remarkable, after the Fish Canyon Tuff was extruded, the volcano continued to be active with at least seven other major eruptions over a span of 1.5 million years.

The reason for the Mid-Tertiary Ignimbrite Flare-Up remains a geologic debate, but most theories couple the volcanism with tectonics along the west coast of North America. The Laramide Orogeny, which resulted in the uplift of much of the Rocky Mountains from Canada to New Mexico, is thought to be related to the subduction of the Farallon oceanic plate beneath North America. The Farallon plate was quite young geologically and was thus buoyant. This likely resulted in a shallow angle of subduction, which caused an uplift of the entire western United States. About 35 million years ago, the last bit of the Farallon plate was subducted, resulting in a major reordering of plate tectonics on the western edge of North America. Without subduction, the Farallon plate began to simply sink through the mantle in a process that is known as "slab roll-back." This allowed the very hot mantle to rise and melt large regions of the lowermost crust, creating the magma sources for the ignimbrites. The fact that the source of the magmas was in the lowermost crust is significant for the composition of the erupting magmas and ultimately the mineralization of the San Juans. Despite the extraordinary volume of the eruptive materials, they were almost entirely devoid of hypogene mineral fluids. However, the collapse of the huge calderas throughout the San Juans developed a structural fabric that would localize much younger volcanic activity, which contained mineralized fluids that were deposited in the fractures and faults and gave rise to the great mineral camps.

In the area defined by the San Juan Triangle (Telluride-Ouray-Silverton), there are four collapsed calderas; Uncompahgre, San Juan, Silverton, and Lake City. The first three were formed during the caldera-related volcanism in the time period of 29 to 27 million years ago. The Lake City Caldera formed at the end of the Ignimbrite Flare Up, about 22.5 million years ago (Ellingson, 1996). The geologic record within the San Juan Triangle is quite complex due to the superposition of the calderas and their structural manifestations. The Uncompahgre and San Juan calderas are the oldest; they were active at the same time and collapsed simultaneously with the eruption of a very large ignimbrite sheet. The ring faults associated with the Uncompahgre and San Juan calderas form an oblong structure that is about 45 by 15 kilometers, trending southwest-northeast.

After collapse the Uncompahgre-San Juan volcanic centers remained active, but at about 27.5 million years ago the volcanism focused on what would become the Silverton Caldera (Bove, et al., 2001) in the southwestern extreme of the Uncompahgre-San Juan depression. The Silverton Caldera, which has a diameter of approximately 20 kilometers, is by far the most important structure for localizing mineral deposits within the San Juan Triangle. And most of the famous mines, including the Sunnyside, Gold King, and those in the Red Mountain district, are along Silverton Caldera fractures. Four million years after the Silverton Caldera collapsed (22.5 million years ago), an eruption in the Uncompahgre Caldera resulted in the formation of the Lake City Caldera. The Lake City and Silverton calderas are connected by a series of fractures known as the Eureka Graben (see "E" in figure, below).

The formation of the Lake City Caldera was the last gasp of the Mid-Tertiary Ignimbrite Flare Up. The rich ore deposits in the San Juan Triangle are associated with structures of the Silverton and Lake City calderas, but the deposits were emplaced 5 to 15 million years after the calderas formed. This mineralization is classified as epithermal and is associated with minor episodes of magmatic activity. In general the mineral-laden hydrothermal solutions migrated from the southeast to the northwest. The resulting deposits are base and precious orebodies that contain a wide variety of ore and gangue minerals. The base metal deposits contain mainly galena, sphalerite, and chalcopyrite while the precious metal deposits are mainly native gold. Silver occurs in a suite of exotic minerals that includes tetrahedrite-tennantite, proustite, and pyrargyrite. Gangue minerals include quartz (most common), calcite, pyrite, pyroxmangite, rhodochrosite, fluorite, and barite.

Throughout the Triangle the fractures associated with the Silverton caldera are mineralized. Within the caldera the ring-fractures dominate along with the Eureka graben. Outside the caldera there are swarms of radial fractures trending away from the caldera center. The Idarado and Camp Bird mines, along with the mines in the Sneffels mining district, are contained in northwest-trending fractures radiating from the Silverton Caldera.

The last of the hydrothermal fluids in the Triangle deposited their minerals 10 to 12 million years ago. The San Juan Mountains began a long period of erosion, which carved the final topography that we see today. Particularly profound in shaping the range were periods of alpine glaciation. Over the last two million years, there have been a dozen cycles of glacial advance that have carved deep U-shaped valleys and left the steep cliffs that define the rugged terrain of the San Juans.

Above: Gold and quartz, 2.8 cm tall
San Juan Mountains
Scott Werschky collection, Jeff Scovil photo

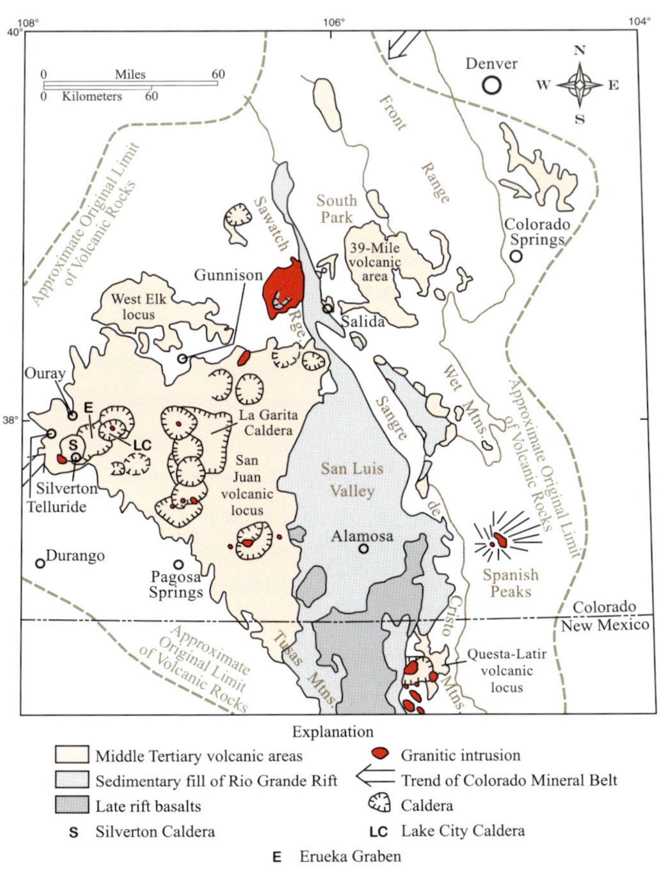

William W. Besse graphic
adapted from Lipman and McIntosh (2008)

Upper: Quartz clusters, 27 cm across, shown *in situ* in a near-surface pocket discovered by Dan and Dianne Kile in the Ouray Amphitheater
Daniel Kile photo

Middle: "The Toll Gate," a mid to late 1880s photo of Otto Mears' famous toll gate near Bear Creek Falls on his newly constructed, 12-mile-long road from Ouray to Red Mountain Town. The road cost $10,000 per mile to build. He ran the toll road for just 8 years before selling it to Ouray County (Tucker, 2003). The original road is the foundation for the modern "Million Dollar Highway."
William Henry Jackson photo
Tom Hughes collection

Lower left: Tom Rosemeyer is pictured here in "the worst hole [he has] had the pleasure of working in"—the 1 level drift near the raise leading to the Richardson stope in the Brooklyn Mine, San Juan County. This stope provided the mine's best modern gold specimens. Note the water rings on the timber posts that show the height of the water in the tunnel before it was rehabilitated.
Robert Stoufer photo, Aug. 31, 1987

Lower right: Ladder to a "Museum pocket," the rhodochrosite crystal-lined vug in the Sunnyside Mine, San Juan County, discussed on page 76.
Jack Murphy photo, summer 1976

Background image: Telluride and the surrounding rugged peaks, as seen from an altitude of 37,000 feet.
Photo Jack Brauer/WideRange Photo LLC

Principle Mines and Specimen-Producing Localities in the San Juan Triangle

Map by William W. Besse

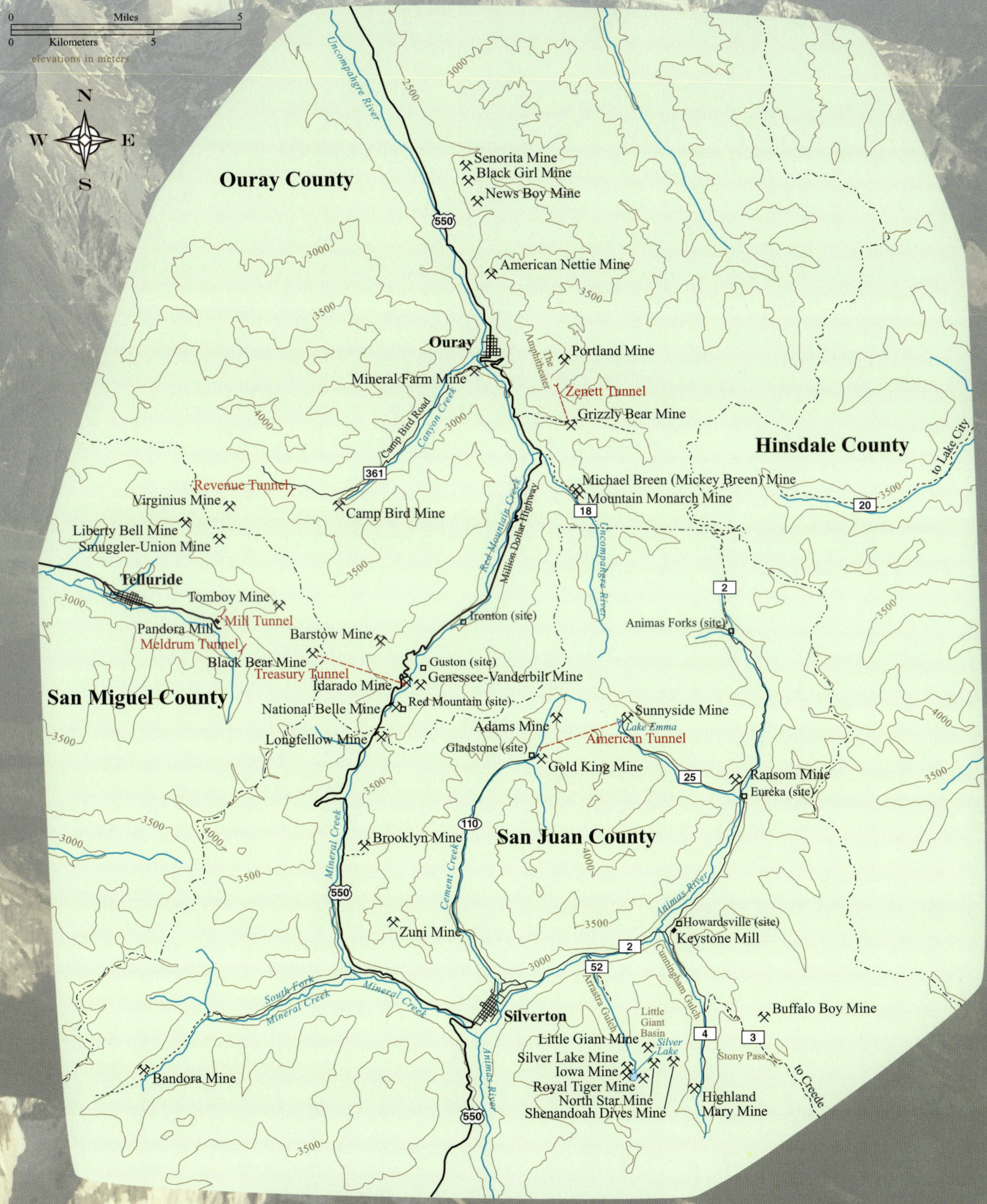

Mines and Minerals along the Rainbow Route: The Red Mountain District

by Mark A. Vendl and Karen A. Vendl

A trip from Ouray to Silverton takes one through some of the most magnificent mountain scenery in the United States and traverses the Red Mountain mining district, which is one of the most interesting and easily accessible mining areas in Colorado. The Red Mountain district is located in the heart of the San Juan Triangle near Red Mountain Pass and is seen by thousands of tourists each summer as they travel the famed Million Dollar Highway between the two towns.

The Red Mountains (there are actually three peaks) are so named because of the characteristic reds, browns, oranges and yellows that come from the oxidation of pyrite in the volcanic rocks, which make a beautiful backdrop for the district. The iron oxide in the mountains makes them appear to glow after a rain. The "Rainbow Route," the nickname for the path of Otto Mears' Silverton Railroad, traveled between Silverton and Ironton, on the north side of Red Mountain Pass. According to Smith (2004), "The wealth of scenery along the way as well as the pots of gold (unfortunately mainly silver) at the mines in the Red Mountain mining district were good reasons for the name."

By the end of the 1870s, most of the mineral producing areas of the San Juans had been located, but prospecting in the area around Red Mountain lagged. A few claims had been staked, but nothing of much importance. Heavy snows limited prospectors, who were further hindered by the fact that the geology of the Red Mountains is different from that of the other districts in the San Juans. Most prospectors were looking for vein deposits, not the "chimney" or breccia pipe bodies, which contained Red Mountain's rich silver-lead and silver-copper ores. These are "nearly vertical bodies of ore roughly circular or elliptical in plan" (Ransome, 1901).

Indeed, as part of the larger San Juan Volcanic Field, the volcanic rocks in the Red Mountains were subjected to intense magmatic and hydrothermal activity that created vertical chimneys or breccia pipes along the ring fault zone of the Silverton Caldera. Mineral-laden solutions deposited the

Above: A Sept. 17, 1888 photo has "Pathfinder" Otto Mears standing in front of the first locomotive, an overhauled Baldwin, purchased by the Silverton Railroad. Mears named this favorite engine "Ouray" and gave it the number 100 (Tucker, 2003) Charles Goodwin photo, Thomas McKee collection

Facing page: Red Mountain #1 makes a glorious backdrop for autumn colors at Ironton Park, along Red Mountain Pass Photo Jack Brauer/WideRange Photo LLC, 2008

rich ores into these vertical spaces or they leached and replaced silicified or argillized rocks (Holmes and Kennedy, 1983). Caves are common in the Red Mountain District, and many are lined with ore minerals such as enargite, sphalerite, proustite, pyrargyrite, galena, covellite, chalcocite, bornite, tennantite, and gold. Heavy sulfide ores may extend hundreds of meters beneath the surface (Fetchenhier, 1996).

In the 1880s, people had pinned their hopes on Red Mountain District to be a second Leadville. But even though the ore in the district was incredibly rich, the orebodies were limited; thus, it never reached the status of Leadville. Most of the Red Mountain mines closed in the 1890s, casualties of the 1893 Silver Crash not to mention the increasingly corrosive water that was being produced as the mines went deeper. Through the years, the district has provided collector quality specimens of enargite, sphalerite, barite, quartz, fluorite, pyargyrite, acanthite, covellite, and tetrahedrite.

The Red Mountain mining district is easily toured by driving the Million Dollar Highway (US 550) from Ouray to Silverton. Heading south from Ouray, travelers encounter several switchbacks before entering Uncompaghre Canyon. The scenery along the next six miles of narrow curving road is spectacular. The road then passes through a long, flat valley known as Ironton Park. The remains of Ironton are at the south end of the park and the northern end of the Red Mountain mining district. In the mid-1880s, Ironton was the largest town in the Red Mountain District. It had twelve saloons and four restaurants along its mile-long main street. In 1889, Ironton became the northern passenger terminus of the Silverton Railroad. Milton Larson was the town's last resident; he died in 1964. Today only a half dozen structures remain (Smith, 2004).

South of Ironton and east of the highway, the **Silver Belle** deposit was discovered in 1880 and was at its most productive between 1887 and 1890. The mine reached a depth of about 700 feet with 10 levels; its ore carried from 500 to 1,000 ounces of silver per ton. The grade of the ore decreased with depth and at the lowest levels of the mine was chiefly composed of pyrite, assaying at less than 20 ounces of silver per ton. Water inflow also increased to as much as 65 gallons per minute with depth. The Silver Belle closed in 1894, due to lower silver prices (Ransome, 1901).

A couple of switchbacks past the Silver Belle, Ouray County Road 31 takes off to the east, crosses Red Mountain Creek, and passes the **Guston Mine**. This deposit was discovered in 1881 by August Dietlat, Andrew Meldrum, John Robinson, and Albert Lang. It was originally mined for lead to be used as flux at local smelters. The mine was sold in 1883 for $100,000. Large-scale development began in 1888, and the mine closed in 1897 when it bottomed in pyrite at a depth of some 1,300 feet. The Guston produced an estimated six million dollars of ore (Smith, 2004). Some stromeyerite ore from the mine assayed at 15,000 ounces of silver per ton (Holmes and Kennedy 1983).

Just a quarter mile south of the Guston is one of the most striking and eye-catching head frames in the district, that of the **Yankee Girl Mine**. John Robinson discovered the Yankee Girl on August 14, 1882, while he was hunting for meat. Stopping to rest, he noticed a rock that was heavier than normal. Upon breaking it open, he found that the rock contained silver-bearing galena. Robinson located the source of the galena and then filed claims for himself and his three partners in the Guston Mine. The partners sunk a twelve-foot shaft and located the first of the rich Red Mountain chimney ore deposits. The first ore from the Yankee Girl Mine contained an average of 88 ounces of silver per ton and was

Left: William Henry Jackson circa 1885 photo of the Yankee Girl Mine
Dave Bunk collection

Middle: Acanthite (low temperature polymorph of argentite) crystals 2 cm long
Yankee Girl Mine
collected by Maynard Bixby
Field Museum specimen (FM8232)
Mark Vendl photo

Lower: Covellite with barite, 8.5 cm wide
Genessee-Vanderbilt Mine
Dave Bunk collection
Jesse La Plante photo

56 percent lead. A month after discovering the Yankee Girl, the partners sold the mine for $125,000 (Smith, 2007). The deposit was mined until 1896 and became one of the richest mines in the Red Mountain District, with a total estimated production of between 3 and 12 million dollars (Holmes and Kennedy, 1983).

Famed Utah mineral collector Maynard Bixby collected in the San Juans during the 1890s. In his 1894 *The Mineral Collector* article, "Notable Minerals in Western Mines," he describes a specific acanthite specimen (pictured right) from the Yankee Girl (acanthite specimens exhibiting cubic symmetry were commonly referred to as argentite):

> *The Yankee Girl mine at Red Mt., Colo., produces exceedingly unique specimens of argentite. These crystals start from the matrix in a prismatic form similar to acanthite, and terminate in an irregular isometric crystal of argentite. A specimen collected by the writer shows a perfect crystal seemingly orthorhombic in habit, in size about ¾ in. x ⅜ in., and very similar on form to crystals of arsenopyrite.*

In 1905 this specimen along with Bixby's entire personal collection was sold to the Field Museum in Chicago, where the specimen resides today.

Across the valley to the west is the **Barstow Mine**. Often included in the Red Mountain district, the Barstow Mine deposits occur not in chimneys but rather in veins. Gold, silver, and fluorspar were the major ore minerals produced at the Barstow. In the early 1900s, the mine produced large specimens of octahedral green fluorite and quartz (see page 58). In the 1930s, the Barstow was incorporated into the Idarado Mine (Holmes and Kennedy, 1983).

The **Genessee-Vanderbilt Mine** is a quarter mile south of the Yankee Girl. Both the Genessee and Vanderbilt deposits were located in 1882. The Vanderbilt was one of the deposits in the district that had a silicious knob on top of the pipe. The lead-silver ore in these mines consisted of galena, and the copper-silver ore was chalcopyrite and tetrahedrite all of which assayed 15 to 40 ounces of silver per ton (Ransome, 1901). The mines were consolidated in 1889 when it was

discovered they were both mining the same orebody. According to Ransome (1901),

> *Cosalite is recorded as having been an important constituent of the ore. Some small caves containing stalactites of pyrite occurred in the ore bodies, but it is not known at what depth. Specimens of ore said to have come from this mine and preserved in the mine office showed some very argentiferous bornite, a slightly argentiferous sulpho-bismuthite of lead containing a little copper and corresponding in physical properties and appearance to*

Upper right: Enargite with malachite coating, 12.2 cm wide
Chimney orebody, National Belle Mine (mined 1880s)
Tom Rosemeyer collection, Mark Vendl photo

Upper left: Barite, 3.5 cm tall, National Belle Mine
Bill and Diana Dameron collection (self collected)
Bill Dameron photo

Lower left: Covellite, 2.5 mm wide, National Belle Mine
Tom Rosemeyer collection, Dan Behnke photo (13.6x)

Below: Quartz after barite on fluorite, 9.8 cm wide
Barstow Mine
Colorado School of Mines collection, Jeff Scovil photo

cosalite, and some small crystals of ruby silver (proustite).

The mine closed in 1897 but reopened in the 1940s and was worked on and off until the 1970s. A few buildings still stand (Smith, 2004). In terms of mineral specimens, individual covellite crystals up to 2.5 centimeters across and groups of covellite up to 5.5 centimeters wide are attributed to the Genesse-Vanderbilt Mine (Eckel, 1997).

Further south and almost at the top of Red Mountain Pass, the **National Belle Mine** is within the city limits of Red Mountain. The National Belle was another incredibly rich mine in the district. It tapped a large siliceous knob, which turned out to be another chimney deposit. While trying to determine the extent of the orebody by driving a crosscut to the main tunnel in July 1883, miners broke into a large cavern. Ernest Ingersoll (1883) described the discovery:

> *This occurred by the accident of a workman breaking through the tunnel wall into a cavity. Hollow echoes came back from the blows of his pick, and stones thrown were heard to roll a long distance. Taking a candle, one of the men descended and found himself in an immense natural chamber, the flickering rays of the light showing him the vaulted roof far above, seamed with bright streaks of galena and interspersed with masses of soft carbonates, chlorides and pure white talc. On different sides of the remarkable chamber were small openings leading to other rooms or chambers, showing the same rich formation. Returning from this brief reconnoisance* [sic] *a party began regular exploration. They crept through the narrow opening into an immense natural tunnel running above and across the route of their working drift for a hundred feet or more, in which they clambered over great bowlders* [sic] *of pure galena, and mounds of soft gray carbonates, while the walls and roof showed themselves a solid mass of chloride and carbonate ores of silver. Returning to the starting point they passed through another narrow tunnel of solid and glittering galena for a distance of forty feet, and found indications of other large passages and chambers beyond. "It would seem," cries the local editor in his account of this romantic disclosure, as though Nature had gathered her choice treasures from her inexhaustible storehouse, and wrought these tunnels, natural stopping places and chambers, studded with glittering crystals and bright mineral to dazzle the eyes of man in after ages, and lure him on to other treasures hidden deeper in the bowels of the earth ... The news of the discovery spread like wildfire, and crowds came to see the sight, and to many of them it was one never to be forgotten.*
>
> *This was only the first of these surprises, for many cavities have since been divulged, great and small, in each of which*

Left:
Fluorite, enargite and quartz
7.7 cm wide
Longfellow Mine
Red Mountain District
Ouray County
Colorado School of Mines collection
Jeff Scovil photo

Facing page, upper left:
Enargite, 5.25 cm tall
Collector's Edge, Inc. specimen
Jesse La Plante photo

Facing page, lower left:
Galena and sphalerite
6.6 cm tall
Longfellow Mine
Red Mountain Pass
Ouray County
Colorado School of Mines collection
Jeff Scovil photo

crude wealth had been locked up since the world was made. The character of the ores, the occurrence of these cavities, and the extremely short distances beneath the turf at which rich ore was struck, have combined to cause much discussion among geologists as to the true history of the district.

News of the discovery spread quickly, and the public was allowed to view the wonder (Smith, 2007). Even larger chambers were soon discovered. The National Belle operated until 1897 and produced silver valued between one and two million dollars. In some of the chambers, radiating prisms of enargite crystals with malachite and quartz overgrowths were found in unattached masses (Holmes and Kennedy, 1983). According to Smith and Rosemeyer (1986), the National Belle Mine produced some excellent crystallized mineral specimens, but few have survived. One that did survive is a 20-by-18-centimeter specimen covered with radiating prisms of enargite (4 centimeters), which are coated with quartz. This specimen is part of the Maynard Bixby collection at the Field Museum in Chicago. Micromineral crystals (enargite, sphalerite, pyrargyrite, and tetrahedrite, to name a few) have been collected in recent years from the mine dumps. Most of this material came from the sulfide zone (Smith and Rosemeyer, 1986).

In the summer of 1994, the Idarado Mining Company began reclamation work at the National Belle. "The work involved diverting the east branch of Red Mountain Creek away from the bottom of the waste dump. The stream was moved about 30 feet to the east; in the process, the ballast from the old railroad grade to the mine was bulldozed up onto the existing National Belle dump. The material was some of the first low-grade ore mined in the 1880s and consisted in part of pyrite, enargite, and quartz. Upon breaking up the ballast revealed beautiful crystallized pockets of quartz, enargite, and associated minerals" (Rosemeyer 1995). Indigo-blue, hexagonal covellite crystal plates up to 2 millimeters on edge were among the microminerals found at this time.

One of the last deposits discovered in the Red Mountain mining district, the **Longfellow**, was just south of the National Belle. The Longfellow, near the southern end of the Red Mountain district, was discovered in 1953 when a chimney was exposed during construction on Highway 550. Claim owner Maud Kemmerling hired mining engineer Warren Prosser to work the deposit. In 1954 Prosser sunk a shaft to a depth of 150 feet. Bill Calhoon who was working in the mine in 1956 tells of blasting into a small (about 3-by-12-foot) chimney that looked like a jewelry store. "The vug was masses of ⅛-inch drusy quartz with brilliant black enargite perched on the quartz and all covered with sparkling white

THE FAIRY CAVE.

"The Fairy Cave" from Ernest Ingersoll's 1890, *The Silver Caves*
Mark Vendl collection

kaolin clay—it looked like Christmas in summertime." Fifteen powder boxes of enargite and quartz specimens were collected by Bill and his partner, and Warren Prosser sold some additional specimens to a Denver mineral dealer. The mine was closed in 1957 and was allowed to fill with water (Tom Rosemeyer, unpublished manuscript).

The mine reopened in 1958 under a lease to Haliburton Petroleum Company. Warren Prosser managed the mine. According to John Crim, who worked in the mine, a chimney with a 6-foot silicious zone was hit while drifting. The center ore zone was about 18 inches in diameter and was full of druzy quartz with enargite and tetrahedrite crystals. The best piece that John collected had 3.1-centimeter enargite crystals on it. He later sold the specimen to Ed McDole (see page 90), who bought specimens from Warren Prosser as well as from the miners. Mineral dealer Benjy Kuehling (see page 42) would travel from Denver to purchase crystals from Warren Prosser when there were enough specimens to make the trip worthwhile. Most of the specimens came from the north orebody (Tom Rosemeyer, unpublished manuscript).

Carl Dismant Sr. and Carl Jr. worked the mine in 1966, but they found little ore left in the mine, and only a few crystals were found (Carl Dismant Jr., pers. comm., 2010). They let the mine flood in December of that year. John Crim and his partner were the last people in the mine before it flooded. John recalled his last few hours in the Longfellow:

> *I knew there were some nice enargite crystals left in the small chimney we found in 1959. So we came back in a couple of hours after we had pulled the pumps on day shift. My partner and I climbed down the shaft and went back to the chimney. The small raise on the chimney was up about 15 feet from the back of the drift. I got part way up the raise on some stulls and started to loosen crystal groups with a scaling bar. My partner was catching the crystals and groups with a fish net. The largest crystals were about an inch long and terminated (enargite). We collected about 60 pounds of specimens and decided to get out. The water was up to our shoulders when we got to the shaft station. That was cutting it a little too close.* (Tom Rosemeyer, unpublished manuscript)

Continuing south from the Longfellow, County Road 31 connects back to the Million Dollar Highway (US 550). From the Red Mountain district, Highway 550 essentially follows the route of the Silverton Railroad to the town of Silverton. The underground workings of the Red Mountain district are inaccessible now, but the mine dumps still yield interesting specimens, mainly microminerals.

Right:
Underground in the American Tunnel
of the Gold King Mine
San Juan County
Tom Rosemeyer collection

Facing page:
Leaf gold on quartz
4.6 cm tall
Smuggler Union Mine, Marshall Basin,
San Miguel County
Colorado School of Mines collection
(part of the Allison/Boettcher
collection)
Jeff Scovil photo

The Golden San Juans

by Robert B. Cook

The major mining districts in the San Juan Triangle (Silverton-Ouray-Telluride) are historically well known sources of many fine mineral specimens, including gold. From the first discoveries in the 1860s, many of the early mines in the western San Juans were known primarily for their silver production. However, significant gold discoveries were not to come until the mid-1890s. During the years of gold discoveries, railroads reached many formerly isolated areas and eased transportation costs for communities such as Telluride, Ouray, Silverton, Gladstone, and Red Mountain. Dependable and relatively inexpensive ore shipments made mining profitable even for some of the smaller mines.

Soon geologists and mining engineers were studying the deposits, mining companies were constructing mine plants and mills, and local smelters were extracting metals from the ores of the region. Coupled with this progressive development was a growing body of knowledge concerning the occurrence of potentially economic gold in many of the already established deposits, and by the beginning of the twentieth century, efforts were increasingly focused on that commodity. In all, approximately 12 million ounces of gold were produced from the San Juans by the time the major mines had permanently closed in the late twentieth century. Surprisingly, however, despite gold's wide distribution and locally very high-grade occurrences, only a relatively small number of specimens seem to have survived the mill. These can be found in regionally focused private collections, major museums, and occasionally in dealer's inventories.

San Juan County

Specimen gold is well known from the **Sunnyside Mine** of the Eureka District (Rosemeyer, 1988). The nearby **Gold King Mine**, located about a mile to the southwest in the Gladstone District, was an important gold producer in the late nineteenth and early twentieth centuries, but the mine was not known for gold specimens. In 1959 Standard Metals Corporation acquired a large group of claims that included the Sunnyside, Gold King, and Shenandoah-Dives, consolidating them into a new mining operation. After extending the Gold King's old American Tunnel over a mile to a depth of 600 feet below the deepest levels of the Sunnyside, they began to mine what was essentially lead-zinc ore on a break-even basis. It was not until 10 years later that the company discovered that a northeast-trending concealed vein (a vein that does not crop out at the surface) carried what was then thought to be an isolated zone with significant gold values; in 1973 core drilling revealed the vein's richness along its length. Soon, the Sunnyside was mining a series of the northeast-trending rich gold veins such as the Little Mary, Climax, Portland, and Spur in addition to old standbys like the Belle Creole, Washington, and Sunnyside (Bird, 1986).

By virtue of these new discoveries, the Sunnyside became the largest gold producer in the state. This was also the most specimen-productive period at the mine, until, on June 4, 1978, disaster struck. The waters from Lake Emma broke through 65 feet of overburden into the Spur vein, dumping tens of thousands of liters of water and more than a million

metric tons of mud into the mine. The two-mile-long American Tunnel, some 2,000 feet below the lake, was completely filled with mud. It took two years to clean out the workings, but the recession of 1981-82 forced Standard Metals into bankruptcy. The mine was allowed to continue operating until March 1985, when work was stopped because of mounting losses (Bird, 1986; Rosemeyer, 1988)

Gold specimens produced during these years are rarely labeled correctly because high-grading miners were concerned that too much information on a specimen label could lead the company to their doorstep. "Sunnyside Mine" and "American Tunnel" are the most common localities listed; "Gold King Mine" is rarely seen on labels; "Sunnyside vein" and "Belle Creole vein" are very rare (and highly suspect) attributions. Most specimens consist of white quartz impregnated with fine, often-connected wiry grains, which vary dramatically from sparse to very dense. Slabbed specimens thickly impregnated with thin wire gold and measuring over 50 centimeters across have been seen in private collections. In some instances the gold becomes so fine-grained and thickly disseminated that it gives the host quartz what has been described as a "buttery" appearance.

Other habits include small groups of brilliant, complex microcrystals and crystal aggregates in vein quartz vugs, up to 7 or 8 millimeters in diameter. Rosemeyer (1988) reports larger hackly masses of gold intergrown with quartz. Gold of the former habit is occasionally found with tiny petzite

Gold from the Sunnyside Mine near Silverton San Juan County Colorado

Upper: Gold sheets and wires on quartz 3 cm tall Dave Bunk collection Jesse La Plante photo

Right: Gold with quartz, 7.4 cm wide Julie Hesse collection Jeff Scovil photo

Gold from the Brooklyn Mine, near Chattanooga
San Juan County, Colorado

Left: Gold on sphalerite, 2.8 cm tall
Dave Bunk collection, Jesse La Plante photo

Above: Gold, 1.9 cm wide
Benjy Kuehling collection, Jesse La Plante photo

crystals, and gold pseudomorphs after petzite have been described from the Belle Creole vein. Another habit is represented by coarse flakes, sheets, and leaves intimately associated with galena and/or sphalerite; beautiful specimens of bright gold masses included in galena were found during the mine's most productive years. In the late 1960s, unusual specimens consisting of relatively coarse gold intimately associated with rhodochrosite and pyroxmangite were recovered and sold briefly at mineral shows nationwide. Calaverite has been reported from polished sections of high-grade ore but is unknown in collectible specimens.

Ransome (1901) described the Sunnyside vein as "galena, sphalerite, chalcopyrite, pyrite, tetrahedrite and free gold, in a gangue of quartz and rhodonite [now identified as pyroxmangite] … The best ore, carrying free gold, occurs in irregular lenses … [and] is usually associated with rosin-colored sphalerite and small amounts of lilac fluorite. These are regarded as useful indications, but are not invariably accompanied by free gold." Burbank and Luedke (1969) reported several stages of mineralization, noting that sphalerite associated with late quartz is generally "rosin-colored," while that found with the early base-metal ore is almost black. It bears repeating that free gold is found in the late-stage assemblage.

Standard Metals' Sunnyside consolidation included many nearby claims that were originally operated by other companies. Ransome (1901) reported that gold was associated with fluorite, calcite, and sulfides as well as rhodochrosite and pyroxmangite. He documented specimens of "beautiful dendritic forms made up of thin branching plates of gold embedded in a gangue of quartz with some pale-pink rhodochrosite" from the narrow but persistent vein in the **Golden Fleece Mine**, a few thousand feet up strike to the northeast of the original Sunnyside Mine. He went on to describe the host quartz as quite dark due to micro-inclusions of sulfide minerals and points out that gold occurred in both the unusually dark quartz and the rhodochrosite.

Also northeast of the original Sunnyside Mine, the **Sunnyside Extension Mine** was the source of gold scattered through "masses of spongy quartz" and as "implanted crystals on the faces of quartz crystals in vugs." Interestingly, gold associated with yellow (as opposed to black) sphalerite, molybdenite (originally identified as graphite), native silver, and a bismuth-lead sulfosalt are described from this mine. Good specimens of gold associated with petzite are attributed to the Gold King Mine, but it is uncertain whether such attributions refer to the "original" operation or the Sunnyside consolidation. Other minor native gold occurrences that produced some specimens and were significant enough to be mentioned by Ransome (1901) include the **Sound Democrat** and **Silver Queen** mines of Placer Gulch near the Sunnyside Mine and the **Royal Tiger Mine** in Silver Lake Basin.

The **Brooklyn Mine** in the Chattanooga area, south of the Red Mountain District, operated briefly from 1978 to 1981. Rosemeyer (1989) reported an unusual gold occurrence in which a pod of black resinous sphalerite was held together by wires and leaves of native gold. The largest sphalerite mass was about 70 centimeters wide and was marked by small pockets of rhodochrosite that contained free-growing elongated crystals and leaves of gold, some of which hosted small petzite crystals. Fine gold-sphalerite micromounts were available from this occurrence for a short while. A single galena specimen with included and emerging gold wires was also reported from this find.

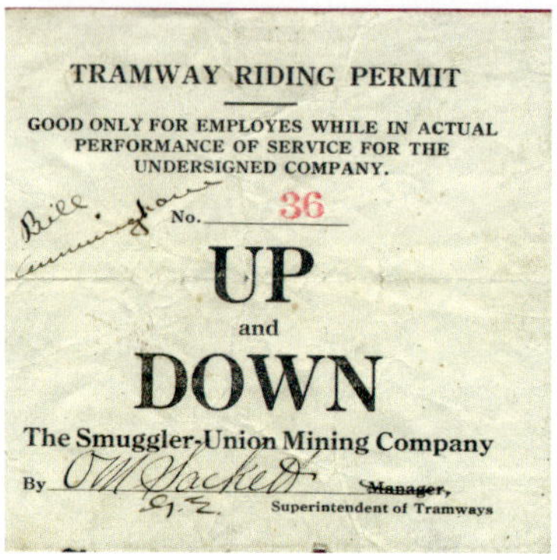

TRAMWAY RIDING PERMIT

GOOD ONLY FOR EMPLOYES WHILE IN ACTUAL PERFORMANCE OF SERVICE FOR THE UNDERSIGNED COMPANY.

No. 36

UP

and

DOWN

The Smuggler-Union Mining Company

By ______ Manager, Superintendent of Tramways

Left: Miners pose in front of the Smuggler-Union Mine boarding house, circa 1910
John Taylor collection

Lower left: Smuggler-Union tram ticket
John Taylor collection

Lower right: Polished section of gold in quartz
field of view 3 cm long
Smuggler-Union Mine
Dave Bunk collection
(ex George Robertson)
Jesse La Plante photo

San Miguel County

The Telluride District is easily the San Juan's most productive gold district. Additionally, this district may well have produced the most aesthetic gold specimens from the San Juans. Unfortunately gold specimens with fraudulent attributions to the Telluride District's mines have plagued the science and hobby for years. Still, many of the major mines are known to have produced fabulous combinations of gold and other minerals.

The Smuggler Union

In 1875, the first important claims (the Sheridan, Union, and Smuggler) were staked on the Smuggler vein in Marshall Basin. Legend has it that J. B. Ingram, J.F. Gundaker, and Jacob Summa discovered a discrepancy in the boundaries of the Sheridan and Union claims and "smuggled" the Smuggler Claim into the space between the two. The Humboldt, Mendota, Cleveland, and Cimarron were staked shortly thereafter, extending the preemption along the vein for 1½ miles. In 1891, the claims along the Smuggler vein were consolidated by the Smuggler Union Consolidated Mining and Milling Company. Operations continued until January 1928, and the properties were idle until purchased by Veta Mines, Inc. in 1936. Following World War II, the Idarado Mining Company purchased the Smuggler Union properties.

The Smuggler Union Mine produced fine gold specimens from time to time. Most are intergrown leaves and ribbons, either free or on matrix. The best of these consist of one or more bright, elongate leaves or ribbons intergrown with, and standing up from, coarsely crystalline quartz matrix. In rare instances, such ribbons reach lengths of 3 centimeters. More common were masses of granular vein quartz held together by massive intergrown hackly gold. According to Purrington (1898), gold occurred in association with acanthite in irregularly shaped grains that exhibited a variety of colors from pale yellow through yellow to red-yellow. Rickard (1907) described crystalline gold aggregates of great beauty and noted wire gold, narrow leaves, or ribbons.

In 1965 several bags of high-grade gold in quartz labeled "Smuggler-Union Mine" were recovered from the basement of an assay shop in Denver. The specimens consisted of somewhat vuggy quartz vein sections that were laced with thin wire and ribbon gold. One of the mine's most famous specimens is a sawn 175.5-ounce quartz mass that boasts a gold vein 10 centimeters wide! The calculated gold content, based on specific gravity determinations performed by the Denver Mint, is 86.64 ounces. This specimen, which won awards at both the 1893 Columbian Exposition and the 1900 Paris Exposition, is now on exhibit at the Denver Museum of Natural History.

Three leaf gold specimens from the Smuggler Union Mine were part of the Allison/Boettcher collection. These were housed at the Colorado School of Mines Geology Museum and were at one time on exhibit in the Arthur Lakes Library. The museum still owns two of these specimens (pictured on pages 22 and 28), and they are currently on display. The third was purchased by the late Richard Kosnar and was illustrated in *The Mineralogical Record* (1983).

The Liberty Bell

In 1876, the Liberty Bell vein was staked in Coronet Basin. The mine produced over 600,000 troy ounces of gold during its 23-year life, but the gold was so finely disseminated in quartz that it was rarely visible to the naked eye (Burbank, 1941).

Above: Gold and quartz, field of view 2 cm wide
Tomboy Mine, San Miguel County
Benjy Kuehling collecton, Jeff Scovil photo

Right: Gold on quartz with rhodochrosite, 12.3 cm tall
Idarado Mine, San Miguel County
Collector's Edge, Inc. specimen, Jeff Scovil photo

The Tomboy

The Tomboy Claim in Savage Basin was staked in 1886 and was soon consolidated with the neighboring Belmont. In the 1890s the two claims underwent several reorganizations, ending with the 1899 formation of the Tomboy Gold Mines Company, Ltd., which was controlled by the London-based, Rothschild-owned, Exploration Company. The Tomboy vein quickly began to play out, and mine manager John Herron exercised an option that he had fortuitously obtained on four claims located on the nearby Argentine vein. In 1911 Tomboy extended its holding northward when it purchased the Montana group of claims on the same vein system from A. E. Reynolds. In 1915 the company purchased the Sydney-White Cloud group still further north from the Montana. In 1927 Tomboy shut down, having produced more than one million ounces of gold and having paid more than four and one half million dollars in dividends.

Like the Smuggler Union, the Tomboy was idle until it was acquired by Veta Mines in 1939. A reorganization in 1940 created Telluride Mines, Inc., and in 1953 the properties passed into the hands of the Idarado Mining Company. Idarado was formed in 1939 with the acquisition of the Black Bear property, which included the southeastern extension of the Montana-Argentine vein. Although the holdings of the Idarado were extensive, the Montana-Argentine-Black Bear vein system was the heart of the company's mining operation.

The Idarado Mining Company focused on the wartime demand for base metals, but gold and silver certainly padded its profits. Production from the operation passed either through the Mill Level Tunnel to the Pandora Mill at Telluride or through the Treasury Tunnel, which connected the southeastern portions of the Black Bear and Argentine veins to a separate mill at Red Mountain Pass in Ouray County. With the Treasury Tunnel portal in Ouray County, many Idarado specimens are labeled "Ouray," but anything from the Idarado complex is probably more precisely labeled "San Miguel County." The Idarado Mine produced a total of 735,000 ounces of gold through 1976.

Purrington (1898) described Tomboy vein gold as spheroidal quartz growths consisting of a core of sugary quartz with finely disseminated gold overgrown with radiating quartz crystals up to 5 centimeters in length. The quartz crystals typically met those from adjacent spheroidal growths, and the entire width of the vein sometimes consisted of these interlocking spherical growths. Although the gold was finely disseminated, it was often visible to the naked eye.

Exceptional gold specimens were found in the Montana-Argentine vein system of the Tomboy Mine. Vuggy areas of the vein contained complex intergrowths of spongy wires, and a few very rich specimens of leafy gold attributed to the Tomboy Mine are in local collections. Argentine vein gold generally exhibits one of four general habits, which include sponge, wire, hackly, and irregular masses or particles between quartz grains. Hillebrand (1957) reported thin plates of gold on quartz crystals and gold of an unusual coppery-red color. Rich, gold-bearing ore produced from the Argentine vein was commonly found in quartz and was associated with abundant galena in some stopes and black sphalerite in others.

After the acquisition of the old Tomboy properties by The Idarado Mining Company, gold specimens began to appear with labels simply stating, "Idarado Mine." But the character of such specimens varies a great deal. Rich leaves and leaf intergrowths, some with exceptional herringbone structure were found. Other specimens consist of white vein quartz

Left: Gold, 8 cm wide
Smuggler Union Mine
Marshall Basin, San Miguel County
Colorado School of Mines
collection (part of the
Allison/Boettcher collection)
Jesse La Plante photo

Below: "Sponge" gold on quartz
3 cm tall
Camp Bird Mine, near Ouray,
Sneffels District, Ouray County
John Trujillo collection
Jesse La Plante photo

Facing page: A circa 1900
photochrome color print of the
American-Nettie Mine in Ouray
Library of Congress Prints and
Photographs collection

with thin ribbons and wires frozen in matrix or emerging into small cavities. A few specimens seen in local collections are bright, dark golden hackly gold in white quartz. It is sometimes possible to make a good guess by comparing specimens to those with better information, but there is an ever-present danger of pushing the identification too far. It is generally best to accept "Idarado Mine" as a sufficiently accurate locality. In fact, gold specimens labeled from the "Tomboy vein" or the "Tomboy Mine" are almost all from the early mining efforts on the Argentine-Montana vein or are products of later mining at the Idarado Mine, as very few specimens were preserved from the Tomboy vein or mine.

The Black Bear Mine

The relatively small Black Bear Mine in Ingram Basin was the source of modest gold specimens in the early twentieth century. The gold occurred in small quartz stringers cutting into the vein's footwall and in greenish quartz near altered fragments of wall rock. Wire gold also occurred in clay slips, and mossy gold masses were found in vuggy white quartz with galena and calcite. Much later, hackly masses of gold, some weighing 16 ounces, were attributed to the Black Bear vein where it was worked through the Idarado Mine.

Ophir Area

Gold specimens were found in the vicinity of Ophir, about 5 miles south of Telluride. Vhay (1962) described gold as occurring in "visible particles, especially in the quartz veins that consist of little else but quartz and gold" and as "wires, more rarely as very thin leaves as much as an inch across, or extremely rarely as large irregular masses." Occurrences include the **Weller**, **Champion**, and **Badger** mines, about 1,000 yards northwest of the original site of Ophir. Purrington's 1896 field notes indicate that gold occurred here as wires in "white vug-like quartz" as well as in calcite. In recent years, reddish sponge and wire gold in white quartz has been collected on mine dumps about 2½ miles east of Ophir and north of the Ophir Pass road.

Ouray County and the Camp Bird

More than two and a half million ounces of gold have been produced from Ouray County, much of it from the famous Camp Bird Mine (see page 30) in the Sneffels District, located 8 miles up Canyon Creek from the town of Ouray. The story of Tom Walsh and his Camp Bird is an important part of Colorado mining lore. In 1896 while prospecting in Imogene Basin for lead ores to use as a flux at his Silverton smelter, Walsh sampled the Gertrude and discovered that it contained rich gold mineralization, with

some assays running $3,000 per ton. After purchasing the Gertrude and adjacent claims, he staked a few of his own, including the Camp Bird. By 1900 Walsh owned 103 claims covering well over 900 acres.

Although not uniformly mineralized throughout its length or with depth, the Camp Bird vein was soon found to host four major ore shoots and later a series of replacement ore bodies at depth. Although a huge producer, the Camp Bird was never a source of fine gold specimens. High-grade ore typically consisted of white quartz with gold disseminated as fine grains or interconnected wiry masses. Some specimens were incredibly rich and consisted of somewhat hackly masses in quartz. Coarse gold wires and leaves in quartz were found toward the end of the mine's life in 1987, during the driving of a raise between the King and Hematite ore shoots. Microcrystals of petzite were occasionally found on the gold (Rosemeyer, 1990).

American Nettie

The Ouray District was an important gold producer with a single significant gold mine: the **American Nettie**. Extraordinarily rich ore is reported from replacement deposits in the Dakota Quartzite by Vanderwilt (1947). Unfortunately, specific descriptions of specimen gold, which must surely have occurred, are lacking.

Hinsdale County and the Golden Wonder

On August 27, 1871, Harry Henson, Jorl K. Mullin, Albert Meade, and Charles Godwin discovered the Ute-Ulay vein system along Henson Creek, on land belonging to the Ute Indians that is west of present day Lake City. A consolidation combined the two discoveries into the Ute-Ulay Mine, which went on to become the most productive in the district by an order of magnitude (Irving & Bancroft, 1911).

The first Lake City mining boom followed on the heels of the signing of the 1873 Brunot Treaty, in which the Utes ceded their land to the federal government (see page 8). In August 1874, Enos Hotchkiss discovered a spectacular outcropping gold vein near the north end of Lake San Cristobal. Originally named the Hotchkiss, the claim was later relocated by George Wilson and Samuel Wendell, who dubbed it the **Golden Fleece**. The gold telluride mineral petzite occurs there in association with lesser amounts of krennerite, calaverite, and sylvanite (Irving & Bancroft, 1911; Slack, 1980; Eckel, 1997). The Golden Fleece is the type locality of hinsdalite.

Similar in mineralization, the **Golden Wonder** vein was discovered in 1880 by Thomas L. Beam near the head of Deadman's Gulch, about 2½ miles south of Lake City. The vein was discontinuous and was worked intermittently by many operators over the ensuing 100 years. In the mid-1980s replacement ore was produced from the F.W. Stub Drift, on the 6 level, that exhibited rich streaks of visible gold with petzite and tetrahedrite when it was cut into slabs. Although geologists periodically commented on the potential of the property, ongoing litigation prevented the mine from being extensively developed until near the end of the 1990s (Brown, 1977; Tom Rosemeyer, verbal communication, 1984). Since then over 100,000 ounces of gold have been produced by a mining partnership, one member of which now owns and operates the famous Colorado Quartz specimen gold mine north of Mariposa, California. Attractive slabs of rich ore containing gold and associated gold tellurides were available to collectors in the mid-2000s. These were from this latest mining effort. Today the Golden Wonder is undergoing intensive exploration by yet another group, and very high-grade intercepts are being reported.

The glory days of mining in the San Juans are past. Although wonderfully productive in terms of total gold production, only a surprisingly small number of gold specimens reliably attributable to the district have survived to the present day. Prospecting via metal detectors has produced precious few specimens relative to the great successes of this new hobby in other areas of Colorado, and it appears that San Juan gold specimens will remain quite rare.

The Camp Bird Mine

by Tom Rosemeyer

Right: At the 14 level Camp Bird mill, freight teams loaded with concentrate to be shipped to the smelter circa 1908 postcard Tom Rosemeyer collection

Facing page: Sphalerite and quartz 6.2 cm tall 285-D stope, Orphan vein 21 level, Camp Bird Mine Ouray County Dave Bunk collection (ex Gary McWilliams) Jeff Scovil photo

Inset: Camp Bird Mine founder Thomas F. Walsh in a circa 1904 G.V. Buck photographic portrait

At the turn of the nineteenth century, the Camp Bird was one of Colorado's most famous gold mines. The mine, located 6 miles southwest of Ouray, operated from 1896 to 1990. In its 94 years of operation, the Camp Bird produced about 1.4 million ounces of gold and 4.6 million ounces of silver. It was also a major source of base metals, producing more than 36,000 tons of lead, 6,000 tons of copper, and 33,000 tons of zinc (Hutchinson 1988).

The mine is accessed from Ouray by following County Highway 361 (Camp Bird Road), which winds its way along Canyon Creek, tracing the route of early trails that were blazed to the high country. The road becomes steep and precarious as it nears the mine and in places has been blasted into cliffs hundreds of feet above the canyon floor. Winters at the mine are long and severe; seasonal snowfall accumulations between 30 and 50 feet are common. Avalanches are a constant danger during the winter months and have taken the lives of many miners along the upper reaches of the road and at the mine.

History

In the 1870s prospectors were swarming the mountains in hopes of striking it rich. A number of rich silver discoveries were made in the region, and the area became known as the "Silvery San Juans." Early prospectors seeking rich silver bonanzas staked claims on a silver-bearing quartz vein high in Imogene Basin, but failing to recognize the rich gold-bearing ore that was contained in the vein, they bypassed a fortune.

It was not until 1896 well after the silver boom had waned, that Irish-born miner Thomas F. Walsh discovered his fabulously rich gold-bearing vein. He named the discovery the Camp Bird Mine after the Gray Jay (*Perisoreus Canadensis*) that was locally called the "Camp Bird" or "Camp Robber." Walsh quickly and quietly acquired almost all of the claims in Imogene Basin and staked all the open ground along the vein. When news of the discovery broke that autumn, there was no rush to Imogene Basin—Thomas Walsh owned everything.

During the next few years Walsh turned his discovery into one of the most profitable gold mines in the San Juans and became a millionaire many times over. Walsh had no partners; he sold no stock; all profits accrued to him alone. By 1900 he was clearing $5,000 per day. In 1902 he sold the mine to London-based firm Camp Bird, Limited. Walsh and his family then moved to Washington, D.C., where he built a mansion and became a member of high society.

Camp Bird, Ltd., mined the deposit for nearly two decades, curtailing operations in 1920. Five years later, Camp Bird, Ltd., leased the mine to King Lease, Inc., which operated the mine until 1956, when the lease was terminated. Camp Bird, Ltd., resumed operations under the name Camp Bird Colorado, Inc. In 1960 the mine was sold to Federal Resources Corporation, which retains ownership today.

The new company mined blocks of ore along the Camp Bird vein. In 1968 base metal replacement ore was discovered at the base of the Telluride Conglomerate, and for the next two years the company concentrated on developing the replacement orebodies, which were mined from 1970 to 1978. Once the replacement bodies were exhausted, the mine was put under the auspices of a caretaker.

In the late 1980s, limited mining resumed on the Camp Bird vein, but that effort only lasted until the spring of 1990 when Federal Resources Corporation closed the mine. Over the ensuing decade, all of the equipment was salvaged and the milling facilities dismantled. Reclamation of the mine site was completed in 2007.

Left:
Galena with calcite, sphalerite, and chalcopyrite
7.8 cm wide
Orphan ROB, 21 level
Tom Rosemeyer collection
Mark A. Vendl photo

Facing page:
Polished slab of mineralized Telluride Conglomerate, 25 cm wide
Orphan ROB, 21 level
Collected by
Tom Rosemeyer
Benjy Kuehling collection
Gloria A. Staebler photo

Geology and Mineralogy of the Ore Deposits

Paleozoic and Mesozoic sedimentary rocks overlain by Cenozoic volcanics are among the rocks exposed in the underground workings of the Camp Bird Mine. The Permian Cutler Formation, consisting of maroon-red interbedded shales, siltstones, and sandstones, is the oldest sedimentary rock exposed. It is found in the deepest level of the mine (21 level). The Tertiary Telluride Conglomerate was deposited on the partially eroded Cutler Formation Unconformity and is approximately 250 feet thick in the mine area. The formation is a gray to red conglomerate of poorly sorted pebbles, cobbles, and boulders of pre-existing sedimentary, igneous, and metamorphic rocks in a sandy matrix. The Telluride Conglomerate is the host rock for the lead-copper-zinc replacement orebodies, which were mined on the 21 level from 1970 to 1978.

The Tertiary San Juan Tuff was deposited on the eroded surface of the Telluride Conglomerate; the tuff reaches a thickness of some 2,300 feet in the mine area. The formation consists of rhyodacite tuff breccias with quartz-latite ash-flow tuffs (Larsen and Cross, 1956). The San Juan Tuff hosts the gold-quartz-base metal ore shoots along the Camp Bird vein as well as the orebodies in the northwest-trending veins. Overlying the San Juan Tuff is the Silverton Volcanic Group, which is composed of the Picayune, Burns, Henson, and Potosi formations. The ore shoots along the Camp Bird vein only penetrate a short distance into these volcanics.

Three distinct types of ore deposits have been mined at the Camp Bird. Although all are epithermal, they differ in structure and mineralogy: 1) gold-quartz-base metal ore shoots occur along the east-west striking Camp Bird vein; 2) northwest-trending quartz veins hosting silver and base metals are located on the north (footwall) side of the Camp Bird vein; 3) lead-copper-zinc replacement orebodies occur at the base of the Telluride Conglomerate and are adjacent to the steeply dipping northwest-trending quartz veins.

Camp Bird Vein

The Camp Bird vein strikes approximately N 85° W, dips 70° to the south, and diagonally cuts across the northwest-trending swarm of radiating veins and dikes emanating from the Silverton Caldera and terminating in the Stony Mountain stock. Four main stages of mineralization evidence several reactivations of the Camp Bird fracture: 1) quartz-sulfide ores; 2) quartz-hematite-gold ores; 3) gold-quartz breccias; and 4) barren quartz stage. Six ore shoots have been mined along the vein. They are, from east to west, the King, Hematite, Gertrude, Bluebird, Discovery, and 3680. The Gertrude, Bluebird, and Discovery were the richest and accounted for most of the gold production.

The vein consists of a gangue of mainly quartz, calcite, and pyrite. Ore minerals include native gold, galena, sphalerite, and chalcopyrite. Other minerals include epidote, fluorite, gypsum, hematite, rhodonite, and scheelite.

Silver- and Base Metal-Bearing Quartz Veins

Numerous northwest-trending radial fissures and tension fractures were formed by collapse of the Silverton Caldera and intrusion of the Stony Mountain Stock. Underground mapping in the Camp Bird Mine clearly shows that some of the northwest-trending veins cut the Camp Bird vein while others terminated in the hanging wall of the vein, indicating that the northwest-trending veins probably formed after the Camp Bird structure. Ore-bearing hydrothermal solutions deposited metallic sulfides and gangue minerals in some of the fissures. Similarities in mineralogy indicate that this deposition coincided with the first stage of mineralization of the Camp Bird vein; it contains no gold values. The main ore minerals are galena, sphalerite, chalcopyrite, tetrahedrite, and rarely hübnerite; gangue minerals include quartz, calcite, fluorite, and pyrite. There has only been limited mining of these veins on the Camp Bird property.

Base Metal Replacement Orebodies (ROB)

Significant Middle Tertiary lead-copper-zinc replacement orebodies occurred at the base of the Telluride Conglomerate, adjacent to the northwest-trending Orphan and Gordon veins, neither of which crop out on the surface. The mineralization does not extend into the underlying Cutler Formation. The replacement ore is characterized by sulfide and gangue minerals that have taken the place of the sandy matrix, pebbles, cobbles, and boulders of the conglomerate.

The Telluride Conglomerate on both sides of the quartz-sulfide vein has been intensely altered by hydrothermal solutions emanating from the vein and formed an alteration halo and replacement orebody within the conglomerate.

Based on the presence or absence of green epidote and manganese-silicate minerals and on the degree of alteration of the original conglomerate, three zones of alteration have been mapped. Zone 1 is characterized by massive sulfide replacement (sphalerite, galena, chalcopyrite, and pyrite) of the clasts and matrix of the conglomerate. Green epidote and quartz are abundant with lesser amounts of carbonates, chlorite, and sericite. Zone 2 consists of disseminated sulfides in the matrix and partial replacement of clasts. The altered conglomerate contains manganese-silicate minerals, chlorite, carbonates, rhodonite, and clay minerals that give the rock a buff to pink color. Zone 3 consists of hydrothermally bleached conglomerate that becomes weaker farther away from the ore zone and finally grades into unaltered conglomerate. The only sulfide present is pyrite that forms small grains in the matrix.

The Orphan vein, so named because it dips to the southwest while the other northwest-trending veins dip to the northeast, was the feeder vein for the Orphan and Walsh replacement orebodies. In drifting along the Orphan vein and replacement orebody, a 15-foot-thick andesite dike was encountered that had a pinching effect on the Orphan vein, which narrowed from 4 feet in the conglomerate to about 8 inches in the andesite dike. On the far side of the dike, the Orphan vein widened to 8 feet and was composed of solid sulfide. On this far side of the dike, the orebody was referred to as the Walsh replacement orebody. Both the replacement orebody and the Orphan vein were mined for a strike length of 4,400 feet and produced 295,000 tons of sulfide ore.

The Gordon vein is parallel to and about 250 feet southwest of the Orphan vein. It was discovered by diamond drilling and was a blind vein with no outcrop and hence had no name. To remedy that situation, the names of all of the Camp Bird miners were written on slips of paper and placed in a hard hat. The office secretary drew the name Gordon Wilbur, a trammer, and thus the Gordon vein was named.

The three Gordon replacement orebodies were not continuous along the vein, but their mineralogy was similar to that of the Orphan and Walsh orebodies. The combined strike length of the Gordon replacement orebodies was 1,350 feet along the strike of the vein; the Gordon replacement orebody accounted for 89,000 tons of ore.

Specimen Mineralogy and Collecting History

In the early years of operation, the Camp Bird Mine was more famous as a gold mine than as a source of mineral specimens. Mining was confined to ore shoots along the Camp Bird vein, which produced few crystallized mineral specimens. In the 1950s and 1960s, miners were collecting two types of quartz. The first were small groups of thin, water-clear crystals, which occurred with the Stage 3 gold-quartz breccia. These were fairly commonly and miners brought them home to place on their windows sills and sometimes to sell to local shops. The second type were large milky-white crystals, the last stage of mineralization in the barren quartz veins. Large, crystal-bearing vugs were encountered that contained milky quartz crystals up to 24 centimeters long. Many of these quartz "points" were broken off by miners and brought home for rock gardens.

In those years, miners were primarily after native gold. The bonanza gold ore that Thomas Walsh mined averaged 1.6 ounces of gold per ton, with some of the ore running as high as 200 ounces per ton. The richest ore occurred in the upper levels of the mine and decreased with depth. All of the gold in the vein occurred in its native state and was alloyed with 5 to 20 percent silver. Most of the gold was not readily visible to the naked eye as it occurred as fine flecks and narrow dark bands in quartz. In some areas along the vein, gold was concentrated as hackly masses in quartz. The gold in quartz was consistently high-graded and almost never made it down the ore pass. Cabinet specimens of gold are scarce as most was "worked down" and sold. Many of the buyers ran legitimate businesses in Ouray but also had small labs set up in their basements to refine crude gold ore.

The gold specimens that have survived were kept rough or were cut into cabochons. The free growing crystallized gold

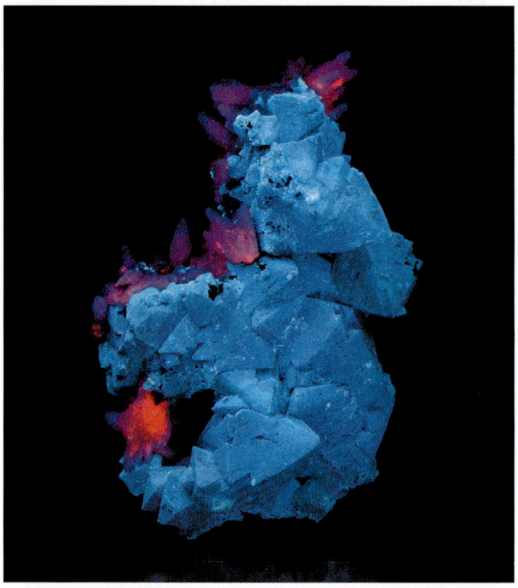

Right: Quartz, 13 cm tall
14-1550 stope
Discovery oreshoot, Camp Bird vein
Benjy Kuehling collection, Jeff Scovil photo

Facing page, upper left and **middle right:**
Calcite on scheelite, 8 cm tall
under incandescent (**left**) and shortwave ultraviolet light (**right**)
5-7850 stope, King oreshoot
Camp Bird vein
George Witters specimen
Both photos by Jesse La Plante

Facing page, upper right:
Crystallized gold in quartz, 5.2 cm wide
349 stope, 3 level, Discovery oreshoot
Camp Bird vein
Mined in the 1930s
Tom Rosemeyer collection, Mark A. Vendl photo

Facing page, lower right:
Mine engineers surveying on the 4 level of the Camp Bird Mine
circa 1908 postcard
Tom Rosemeyer collection

Facing page, lower left:
Calcite on dolomite after calcite
13 cm tall
Walsh ROB, 21 level
Colorado School of Mines collection
Jesse La Plante photo

occurred in tiny quartz vugs and usually formed wires or small groups of crystals perched on quartz. The Camp Bird vein also contained appreciable amounts of sphalerite, galena, and chalcopyrite, but all generally occurred as a massive vein filling rather than being crystallized in vugs.

In the late 1980s, the Camp Bird vein was mined for scraps of gold left by previous operators. This effort exposed a large vug containing hundreds of scheelite crystals. The dipyramidal crystals reach 2 centimeters on edge and range in color from white to dark brown. The crystals are associated with sugary fluorite and are some of the best scheelite specimens recovered from Colorado.

Back in the mid-1960s, mining commenced on base metal (lead-copper-zinc) replacement orebodies along the Argentine vein at the nearby Idarado Mine. Camp Bird Colorado, Inc., followed suit, initiating a diamond drilling program to explore for replacement orebodies on the Camp Bird property. In 1967 and 1968, the 8572 and Orphan veins were drilled from the 14 level haulage tunnel, which intersected both veins. Forty drill holes were made, delineating an orebody approximately 715 feet below the haulage level. Sufficient ore was blocked out to justify sinking a shaft.

The vertical shaft was at the intersection of the 8572 drift and the 14 level haulage tunnel. Construction on the 800-foot shaft and 21 level station began in late 1968 and was completed in early 1970. In July 1970, the outer alteration zone of the replacement orebody was encountered. I was fresh out of the Army and had secured a position as mine engineer at the Camp Bird. When I started work in August 1970, the replacement orebody had just been opened and was proving to be as rich as the diamond drilling had indicated.

The first part of the Orphan replacement orebody that we mined was in a highly mineralized basal sandstone unit at the base of the Telluride Conglomerate and was devoid of crystallized mineral specimens. It was not until the Orphan vein was intersected by crosscuts from the main drift, which was being driven southeast along the edge of the replacement orebody, that we encountered crystallized specimens. The few specimens that were recovered from pockets in the massive sulfide vein, which had an average width of 20 inches, were galena, sphalerite, and chalcopyrite in association with each other or as individual crystals up to one centimeter on quartz crystals, a few centimeters long.

After two crosscuts had intersected the Orphan vein, we determined that the vein had a strike of S 45° E and a near vertical dip. The basal sandstone unit was a narrow ancient river channel, which had been cut into the coarse conglomerate. Once the drift was out of the channel and back into the conglomerate, leached clasts containing crystallized specimens of ore and gangue minerals were found, indicating that the prospects for future specimen collecting were good.

As mining continued in the Orphan replacement orebody, more and more crystallized clasts were encountered—a treasure house of crystallized minerals. The clasts nearest the vein were usually totally replaced by ore minerals but the ore-bearing stage was less intense from 15 to 50 feet from the vein. The best crystal pockets, which ranged from a few centimeters to 25 centimeters, were in this zone. Generally the leached casts were circular or oblong and filled with some of the most beautiful crystallized minerals that I have seen in my 40 years of underground mining. In general, vugs were partly exposed by the fresh blast with its contents either still in place or scattered on the muck pile. The crystallized clasts within the blasted round didn't fare too well and were usually blasted to bits. Most of the crystallized minerals were on the outer edge of the vug or loose crystals in the pockets.

The most common ore mineral in the vugs was sphalerite, which formed lustrous black crystals in singles or groups that ranged up to 6 centimeters. Associated with the sphalerite were sharp, cubic crystals of galena up to a few centimeters on edge. A few brassy chalcopyrite crystals were also collected, but they were scarcer than were the first two sulfides. The gangue minerals formed a substrate for the sulfide minerals and crystallized as crusts on the sphalerite.

I could hardly contain my excitement when I started to clean out a partially replaced boulder that hosted a plethora of well-crystallized minerals. I carefully pulled out specimens of manganocalcite partly encrusted with small sphalerite crystals and pistachio-green, needle-like microcrystals of epidote. I also encountered vugs lined with stubby quartz crystals and rosettes of platy, snow-white barite crystals up to 3 centimeters across. As if that weren't enough, some vugs contained groups of pyrite pyritohedra up to 5 centimeters across floating in a mush of microquartz and epidote crystals.

As the main drift was driven southeast along the Orphan vein, a 13-foot thick andesite dike was encountered. The Orphan vein narrowed from 4 feet in the conglomerate to just 8 inches in the dike. On the other side of the dike, the vein expanded to 7½ feet and was solid sulfide. The replacement orebody also expanded to almost 72 feet wide and was intensely mineralized. The andesite dike had evidently acted as a dam for the hydrothermal solutions traveling northwest from the Camp Bird vein. This area was full of vugs, which were filled with some of the best specimens recovered from the replacement orebody. These included large groups of pyrite, sphalerite, chalcopyrite, and quartz crystals.

From that point on, the orebody was known as the Walsh replacement orebody (in honor of Thomas Walsh), but it was still on the Orphan vein. For the next four years the Walsh replacement orebody was mined along the Orphan vein for a strike distance of 2,500 feet, until the ore became uneconomical to mine. The mineralogy was the same as the Orphan replacement orebody, but the copper content was higher and the number of vugs containing chalcopyrite crystals increased. The Walsh replacement orebody produced the best chalcopyrite crystals, some of which reached lengths of 15 centimeters. The rare mineral digenite was also found coating some of the chalcopyrite crystals, which had formed in the Orphan vein and were later altered by mineralized hot water flowing through the fractured vein. The Walsh replacement orebody contained zones of light green cubic crystals of fluorite associated with crystallized dolomite and siderite. This combination made very aesthetic specimens.

As drifting continued southeast along the Orphan vein and the Camp Bird vein was approached, the grade of ore decreased and the inflow of water increased. As we neared the Camp Bird vein, the open fissure in the face of the drift was pouring 2.5 million gallons of water per day and posed a major pumping problem.

The Gordon #2 replacement orebody was near the end of the Walsh replacement orebody and was of a much lower grade. But the largest and best sphalerite crystals found on 21 level came from the Goron #2. A large vug along the Gordon vein contained some 75 superb groups of large and lustrous black sphalerite crystals. Almost all of the crystal groups were recovered, as the large vug occurred in the back of the stope and were not damaged by blasting. The crystals ranged up to 7 centimeters and occurred as singles and groups.

In retrospect, the golden era of specimen collecting in the Camp Bird mine was not from the gold-bearing Camp Bird vein but from the replacement orebodies, which produced hundreds of first-class specimens that now grace many Colorado mineral collections. I look back at my 40-year career of underground mining and can say that my most enjoyable years were in the 1970s when the Camp Bird, Idarado, and Sunnyside mines were going full blast and pouring out ore and mineral specimens every day.

Right:
Gypsum, 14 cm wide
6 level, East Camp Bird vein
Dave Roter collection
Jesse La Plante photo

Middle left:
Sphalerite, 3.5 cm wide
Gordon ROB, 21 level
Dave Bunk collection
(ex Gary McWilliams)
Jesse La Plante photo

Middle right:
Fluorite on pyrite, 3.5 cm wide
2 SE stope, Orphan ROB, 21 level
Dave Bunk collection
Jesse La Plante photo

Lower left:
Chalcopyrite, sphalerite, quartz
Orphan ROB, 21 level
6.3 cm wide
Dave Bunk collection
Jesse La Plante photo

Lower right:
Calcite, 4.5 cm tall
630-1100 Stope, Walsh ROB, 21 level
Collected in 1974
Tom Rosemeyer collection
Mark A. Vendl photo

Facing page, left:
Calcite and quartz
10.5 cm wide
630-1600 stope, Walsh ROB, 21 level
Collected in 1978
Tom Rosemeyer collection
Mark A. Vendl photo

Facing page, right:
Pyrite, sphalerite and quartz
4.5 cm tall
2 SE stope, Orphan ROB, 21 level
Benjy Kuehling collection
Jesse La Plante photo

Right
"Amethystine" quartz
12.5 cm wide
Idarado Mine
Telluride District
San Miguel County, Colorado
Collector's Edge, Inc.
specimen
Jeff Scovil photo

Facing page:
Quartz, 20.5 cm tall
Manganese Mine
Squaw Gulch
Ouray County, Colorado
collected by Tom Rosemeyer
Collector's Edge, Inc.
specimen
Jeff Scovil photo

Quartz: The Crystal King of the Triangle

by Terry C. Wallace and David W. Bunk

Spectacular examples of crystallized quartz occur throughout the San Juan Triangle: huge milky quartz clusters from the Idarado and Camp Bird (page 30) mines, pale amethyst points from the high mountains around Cunningham Gulch, brassy chalcopyrite crystals on translucent clusters of quartz from the Bandora Mine, and wonderful quartz pseudomorphs after anhydrite (page 44) from pockets in the Leadville Limestone above Ouray are some of the most collectible minerals from the triangle.

Much of the most important ore-bearing mineralization in the San Juans occurred 10 to 12 million years before present, when small intrusions of high-silica alkali rhyolite invaded the region. These intrusions produced multiple episodes of epithermal fluids that in turn produced fracture and fissure filling veins. These veins are controlled structurally by the various calderas in the San Juans; in fact, many of the veins are concentric rings outlining the collapsed calderas. Many of the most important mines in the triangle exploit veins radiating from the concentric rings; the radiating veins formed when the caldera collapsed. This swarm of mineralizing events repeatedly brecciated and rehealed the structures and created large vugs and cavities within veins through the district.

Although the nature of these hydrothermal episodes is localized, many were dominated by quartz. Almost every mine in the district has produced crystallized quartz. Collectible specimens come from every part of the San Juan Triangle, and new specimens come to the market every year.

The Idarado Mine, San Miguel County

The Idarado (see page 46) is one of the dominant mines in the San Juan Triangle, incorporating more than 80 miles of drifts and connecting Red Mountain Pass to the Pandora mill near Telluride. The mine is famous for large, well formed clusters of milky quartz crystals, the largest in excess of 100 pounds. The main veins in the Idarado are wide, quartz and quartz-sulfide deposits filled with vugs and cavities. These vugs, which are typically ellipsoidal, range in size from a few inches in diameter to 25 by 10 feet and are commonly lined with quartz crystals. Much of the quartz from the Idarado Mine is "milky"—white or gray-white. The color is due to minute fluid inclusions or gas bubbles and is thought to indicate pulsing or unstable hydrothermal conditions driving the vein fluids. In fact, studies on the fluid inclusions in a diverse sample of Idarado milky quartz indicate a wide range of emplacement conditions with temperatures varying from 250 to 315 degrees Celsius and a salinity of 0.1 to 7.9 weight percent sodium chloride (Nash,

Left: Circa 1880 view of the North Star Mine high up Cunnignham Gulch on the slopes of King Solomon Mountain Joseph Collier photo Dave Bunk collection

Lower: Clear quartz is relatively unusual in the San Juans. This 8 cm wide specimen was recovered from the 2400 level Basin vein Idarado Mine Telluride District San Miguel County.

John Trujillo collection Jesse La Plante photo

1975). The crystal morphology of the milky quartz shows considerable variability: large crystals are commonly doubly terminated and barrel shaped, the edges curving to a sharp point with no noticeable separation between prism and pyramid.

The Idarado Mine also produces clear quartz crystal groups and pale amethyst. The Denver Museum of Nature and Science houses a doubly terminated translucent pale amethyst crystal from the Idarado that is 18.3 centimeters long. The "Secret vug" contained pale to medium colored amethyst crystals to 48 centimeters long that taper from the base to the tip with only a small pyramid termination. Scepters and "reverse" scepters generally exhibit milky to clear stems while the scepters are commonly gemmier and colorless, although some are amethyst.

Around Cunningham and Arrastra Gulches

Some of the best San Juan quartz specimens come from two drainages that are home to some of the richest mines around Silverton. Arrastra and Cunningham gulches are north-south drainages about one mile apart; Cunningham Gulch is a deep north-south canyon that drains from the south into the Animas River near the Howardsville town site. Between these two Gulches are a series of high peaks—King Solomon Mountain, North Star Peak, and Little Giant Peak—that are home to some of the richest silver mines in the triangle.

The alluvial deposits at the mouth of Cunningham Gulch were some of the original placers discovered by the Baker party in 1860. W.H. Cunningham, a member of that party, was assigned to explore the gulch; hence the name. The region is home to the earliest hard rock mining in the triangle and was the heart of the late nineteenth century Animas mining district. Galena Mountain, Sheep Mountain, and the pathway to Stony Pass along the Continental Divide are on the eastern margin of Cunningham Gulch. Many of the most famous mines in the Silverton area, including the **North Star**, **Highland Mary**, **Shenandoah-Dives**, **Osceola**, **Pride of the West**, **Old Hundred**, and **Green Mountain** all lay within Cunningham Gulch. The richest silver mine in the area, the **Silver Lake Mine** is situated at the head of Arrastra Gulch.

All of the mines in the Cunningham Gulch region exploit epithermal veins with some carbonate replacement. A dominant characteristic of the epithermal veins is the large amount of water they contain. There are also a huge number of quartz veins, which are barren for ore but are prolific producers of quartz crystals. It is still possible to collect crystals along the high mountain peaks; in fact, while climbing King Solomon Peak in the late 1980s, one of the authors (TW) found hundreds of floater pale amethyst crystals to 5 centimeters scattered along the surface at an

Upper left:
Quartz, 6.5 cm tall
Little Giant Basin
San Juan County
Dave Bunk collection
Jesse La Plante photo

Upper right:
Quartz, 11.1 cm wide
Silverton Area
San Juan County
Julie Hesse collection
Jeff Scovil photo

Right:
Quartz, 31 cm wide
Ohio Mine
Ouray County
Dan and Dianne Kile collection
Daniel Kile photo

elevation of about 13,000 feet. Similar stories are not uncommon.

Collector and businessman Robert Stoufer recalls a time in the early 1980s when he was exploring the area just north of Silverton and noticed a quartz outcrop a little way up Cunningham Gulch on a steep slope just off the roadway. He began uncovering clay-filled pockets with groups of centimeter-size light amethyst crystals, which exhibit unusual darker amethyst phantoms. Just below the area that had yielded the amethyst, there was a horizontal zone containing groups of very fine drusy milky quartz pseudomorphs after what has been shown to be anhydrite. This lower zone didn't produce amethyst, but did contain centimeter-size dipyramidal milky quartz crystals, which were intergrown with the pseudomorphs. Stoufer later researched the area's claim maps and found that the quartz zone was on the **Keystone mill site**; the Keystone claim is on the other side of Cunningham Creek.

In fact fine quartz specimens have been attributed to dozens of localities in and around Cunningham and Arrastra gulches. At least three mines, the **Hematite**, **Osceola**, and **Pride of the West**, have produced clear quartz with green chlorite inclusions. Just to the west of Cunningham Gulch in **Little Giant Basin**, a variety of quartz specimens have been found over the last 25 years, including amethyst scepter crystals to 15 centimeters, milky quartz pseudomorphs after anhydrite, and cactus-like growths of tiny milky quartz on larger quartz crystals. Pale amethyst crystals to 10 centimeters have been found along the ridge of **Green Mountain**, between Stony Pass and Cunningham Gulch.

The Indefatigable Benjy Kuehling

by Gloria A. Staebler

Right: Benjy Kuehling working the LJP pocket at the Potomac Mine in San Juan County in August 1998
Tom Rosemeyer photo

For nearly five decades, Benjy Kuehling has been an ambassador of San Juans minerals; his Columbine Mineral Shop in Ouray a required stop for visiting collectors.

Born in Baltimore, Maryland, in 1938, Benjy moved several times with his parents before settling in Denver in 1948. It was there that he met neighbor, minister, and mineral collector Durbin Thorpe, who introduced Benjy to minerals and provided a welcome haven from the turbulent relationship that Benjy had with his father. Durbin employed the boy, paying him in minerals. And in the summer, they went together to the San Juans to buy and collect minerals.

Benjy quit school and left home when he was 16 years old, taking a job as a delivery boy at Riley Reproduction and a $35 per month room in downtown Denver. Drawing on his experience with Durbin, Benjy later went to work for Missouri Minerals and Eckert's Scientific, preparing minerals and managing the counters. Continuing to spend his summers in the San Juans, he bought and collected minerals to sell to the shops in Denver as well as to build his own budding collection. In 1965 when he was just 26 years old, Benjy sold his mineral collection to dealers Merle Reid and George Robertson, bought a building in Ouray, and opened the Columbine Mineral Shop.

For the first several years, the fledgling business was only open in the summers, as Benjy spent his winters working for Missouri Minerals back in Denver. In the mid-1960s, the Idarado, Camp Bird, and Sunnyside mines were all going strong, providing a seemingly endless supply of mineral specimens, so much so that Benjy's friend, the legendary mineral dealer Ed McDole, had taken up residence in Silverton to gain ready access to fresh material.

In 1968 Benjy took his first booth at the Tucson Gem and Mineral Show. By the summer of 1969 at the age of 30, Benjy's business was rolling, and he was thinking about settling down. Enter Chicago-born, Denver schoolteacher Elizabeth "Liz" Spangler, on vacation with a friend in the San Juans. Benjy showed the young women around the area and finally asked Liz if he could take her to a movie when he returned to Denver in the winter. The two dated for two winters and were married in the spring of 1970. Foregoing a honeymoon, Benjy and Liz opened a mineral, jewelry, and gift shop in Aspen and spent that first summer together as they have spent the 41 summers since, in the mountains running their tourist-intensive businesses. Daughter Lauri was born in 1972; son Ben came along in 1974.

Liz and Benjy sold their shop in Aspen to Tom McKee in 1978, but expanded their businesses in Ouray, putting up a new building on Main Street and opening a jewelry store. They also bought (and sold) a number of properties in Ouray and today own 6 buildings in town.

In the late 1990s, Benjy began to scale back. When he set up for the Tucson Show for the last time in 1997, he had served as a dealer at the show for a record 30 consecutive years. Benjy transferred Columbine Mineral Shop to son Ben in 2007; Liz and Benjy own (and Liz still manages) Columbine Gifts, just across the street.

Having sought respite in the pockets of the San Juans since Durbin Thorpe first took him collecting in the 1950s, Benjy has logged thousands of hours in the field and has built an impressive display of self-collected San Juans quartz. He has also "silver-picked" an enviable collection of Colorado minerals. And as if that were not enough, Benjy has fed, but has seemingly not yet satisfied, a museum-size appetite for collectibles including mining artifacts, vintage cars and headlamps, and antique furniture.

A visit to Ouray is incomplete without a chat with the curmudgeonly, straight shooting, good hearted Benjy Kuehling and a glimpse into the house that minerals have built.

Above: Nestled in the San Juans, the town of Ouray was the inspiration for the utopian city Galt's Gulch in Ayn Rand's novel *Atlas Shrugged*. In this photo taken from the west, the May sun is setting on the Amphitheater, the glacial cirque east of town that has drawn mineral collectors for decades.
Photo Jack Brauer/WideRange Photo LLC, 2008

Below: Map of the collecting areas around the town of Ouray, Ouray County
William W. Besse cartography

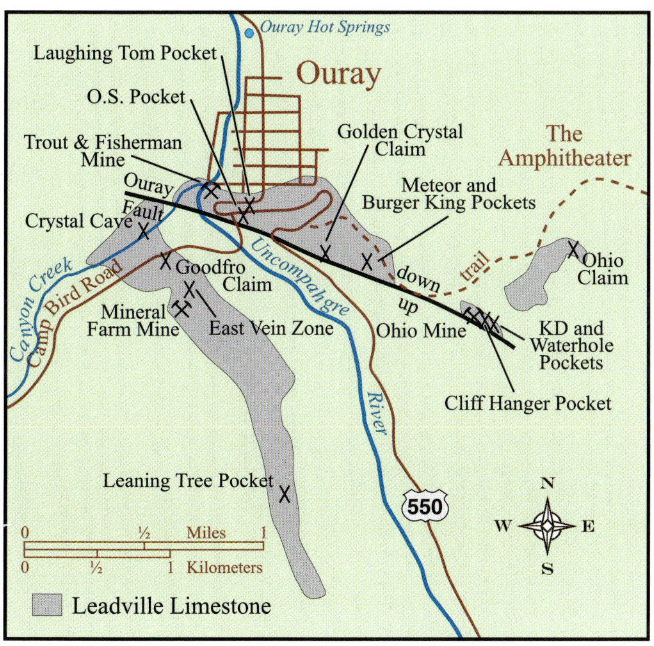

Uncompahgre Mining District, Ouray County

Fabulous quartz specimens have been found around Ouray, and in the last thirty years, many extraordinary specimens have come from the modest Uncompahgre mining district south and east of town. The mines of the Uncompahgre district are mostly in the cliffs and high mountains overlooking Ouray. Unlike the rest of those in the San Juan Triangle, the Uncompahgre mineral deposits are controlled by the ancient structures associated with the Paleozoic-era Uncompahgre uplift. The metals are related to intrusions associated with the Laramide orograny that formed dikes and sills within a thick sequence of sedimentary rocks that span from the Precambrian to the Cenozoic and includes the Leadville Limestone and numerous quartzite units.

The district's most spectacular specimens come from mines and prospects located in a large glacial cirque known as the Amphitheater, a broad bowl due east of Ouray that was formed by glacial carving at the confluence of two Alpine glaciers moving out of the highlands of the San Juans into the Uncompahgre River Valley. There are several mines in the Amphitheater, and all have some commonality. The mines in the Ouray Gold belt, including the **Wanakah**, **American Nettie**, and **Jonathan** exploit small silicified deposits and are largely pyritic gold-bearing ores. Throughout the Uncompahgre district, quartz is a dominant vein mineral; it probably leached and mobilized from the Precambrian quartzite units.

Although quartz specimens have been reported from the Amphitheater for more than one hundred years, the **Portland Mine** and the collector-named **"Ohio Mine"** have in recent years produced remarkable modern specimens. The Portland Mine is high up in the Amphitheater, at the head of Portland Creek. In the 1970s the mine was worked

Sphalerite on quartz with pyrite, 3 cm tall
Portland Mine, Ouray County
Dan and Dianne Kile collection (self collected)
Daniel Kile photo

for mineral specimens, and Kile et al. (1991) describe, "Lustrous, stubby, translucent crystals with milky bases and clear tips" that are generally less than 2.5 centimeters long, but crystals exceeding 10 centimeters in length have been found. Kile et al. also report large crystal groups, and prismatic, gemmy amethyst crystals more than 7 centimeters in length. Some of the quartz specimens from this mine are remarkable for their association with gemmy green sphalerite crystals many of which have a distinctive, light sprinkling of micro-pyrite crystals.

One of the better known modern quartz localities in the San Juans is a prospect located near Ohio Park, known among collectors as the **Ohio Mine** (page 88). The milky quartz from the Ohio is among the world's best: numerous collections are graced with spectacular, lustrous, white, quartz crystal plates to 46 centimeters across with individual crystals to 25 centimeters long or more. Specimens from the Ohio have probably been known since the 1890s, but its modern history is marked by recovery of specimens in the 1980s and 1990s.

The locality is unusual in that the crystals were not collected from individual pockets, but rather from a thick clay zone beneath a region of roof collapse. The quartz originally formed in a large solution cavity within the Leadville Limestone; miners followed an almost vertical quartz vein on one side of and below this cavity. Sometime after the quartz crystals formed, the roof collapsed spreading random clusters of crystals through the solution cavity. The mine was repeatedly inundated over time and deposited a thick sequence of clays. The clay encased the quartz clusters and other debris from the roof collapse. Modern collectors dug through the clay and debris and found amorphous "lumps" that enclosed the quartz crystals.

Quartz Pseudomorphs after ~~Barite~~ Anhydrite from the San Juans

by Virgil W. Lueth

Large tabular quartz pseudomorphs and epimorphs have been collected and sold from the Silver Point Mine and "Oh Shit!" (OS) pocket in the Ouray area of the San Juan Mountains, and for more than a half-century, such pseudomorphs have been identified as quartz after barite. The specimens consist of prismatic drusy quartz that formed over distinctive tabular to blocky crystals. In most samples, no vestiges of former minerals are present in the molds and some are completely filled with quartz. Nineteen individual specimens from the Silver Point (9) and OS pockets (10) were examined to determine the original mineral that is now a mold or has been pseudomorphed by quartz. Comparisons were made to crystal sketches predominantly from Goldschmidt's *Atlas der Kristallformen* (1886–1891). Associated minerals were also noted along with internal textures of the molds. Finally, a survey of the modern thermochemical literature was undertaken to predict the most likely mineralogy of the original crystals.

Muntyan and Muntyan (1988) provided the first written report on these specimens, and they first allude to the possibility that these may be pseudomorphs after anhydrite (also reported by Kile et al., 1991). Their observations are plausible as barite is four orders of magnitude less soluble than

Upper: Quartz pseudomorph after anhydrite, 13.8 cm wide
Silver Point Mine, Ouray County
Benjy Kuehling collection (self collected), Jeff Scovil photo

Lower: Quartz pseudomorph after anhydrite, 15.5 cm tall
OS pocket, Ouray County
Dave Bunk collection, Jesse La Plante photo

Facing page: Interior of a blocky, Silver Point Mine pseudomorph (10 cm long), the pink color is the result of tiny grains of hematite. Cleavage of the original mineral is clearly evident as quartz preferentially replaced the anhydrite along cleavage traces. Most of the quartz overgrowth is outward from the original crystal interface.
Dave Bunk Minerals specimen
John Rahart photo

anhydrite at any temperature. Barite is one of the most insoluble sulfate minerals found at the earth's surface, which makes it one of the most common. Barite displays "prograde solubility" under most conditions, becoming less soluble at lower temperatures and so inhibiting the possibility of dissolution as hydrothermal solutions cool. The "retrograde solubility" of anhydrite (more soluble in colder solutions than in hot) sets up the scenario for later replacement. Anhydrite also alters readily to the mineral gypsum at temperatures below 70 degrees Celsius. Hydrothermal fluid modeling suggests that anhydrite would be fairly common in the early paragenesis of minerals precipitated from the high sulfidation fluids that characterize the ore deposits in the San Juans.

The pseudomorphs from the two localities are significantly different in both habit and development of quartz crystals, yet both provide clues to their original mineralogy. Samples from the OS pocket are more tabular and smaller quartz crystals comprise the druse compared to the Silver Point specimens, which are blocky and have larger quartz crystals. The tabular shapes of the specimens from the OS pocket are quite distinctive. Some display "fish-tail" twinning, a feature observed in anhydrite but not commonly in barite.

The layers of quartz are much thinner than those from the Silver Point Mine, and the internal features of the molds examined are completely obliterated by secondary quartz growth. Large quartz crystals obscure the blocky crystal habit on the exterior of the samples from the Silver Point Mine. However, the textures on the inside of these crystals are magnificently preserved, illustrating the cleavage traces (as thin ledges) and growth habits of the precursor mineral. It appears that the quartz precipitated during the actual dissolution of the anhydrite, concentrated along cleavage traces and crystal boundaries. These textures are not typically seen in barite.

The presence of other minerals in the pseudomorphs is rare. Pyrite and less commonly hematite was observed on the mold-quartz interface in samples from the OS pocket. The samples from the Silver Point Mine often display a stain of hematite on the same interface, suggesting an earlier generation of pyrite that is now oxidized. The presence of pyrite at the interface of the quartz and anhydrite probably indicates a reduction of sulfur during the dissolution of the

anhydrite. Only two samples from the OS pocket contained a very fine grained, white substance that was located on the inside of pseudomorph and inside "holes." X-ray diffraction analysis of this material identified it as gypsum, perhaps a byproduct of the original anhydrite.

The sum of the evidence indicates that these pseudomorphs were originally anhydrite. If the original mineral was barite, it should persist in the specimens and be found inside the molds, a common feature of quartz encrustation pseudomorphs after barite. Anhydrite tends to dissolve readily and would be expected to be "missing" or completely replaced by quartz as is the case in these specimens. The features of twinning, cleavage, and form all suggest the original mineral was anhydrite. The overwhelming differences in solubility between the two minerals is perhaps the strongest evidence. Barite is a common mineral because of its insolubility and anhydrite is comparatively rare for the opposite reason.

Left:
"Treasury Tunnel Mine, Ouray, Colo."
circa 1930s postcard
John Taylor collection

Facing page:
Gold in quartz
3.75 cm wide
Idarado Mine, Telluride
San Miguel County
Benjy Kuehling collection
Jesse La Plante photo

Crystal Collecting at the Idarado Mine

by John Trujillo, Allan Young and Don Earnest

The Idarado Mine is situated in the rugged San Juan Mountains of beautiful southwestern Colorado, known as the "Switzerland of America." Mountain peaks in the area of the mine exceed 14,000 feet in elevation, and winters are frequently long and harsh, with annual snowfalls commonly exceeding 400 inches. The mine is located in the Telluride mining district and underlies parts of two counties, Ouray in the east and San Miguel in the west, with a high mountain divide forming the boundary between the two. The former site of the mine's main office and the portal of one of the its two main adits, the Treasury Tunnel, is situated in Ouray County just north of Red Mountain pass, at an elevation of 10,600 feet. It is accessed via U.S. Highway 550, commonly referred to as the "Million Dollar Highway." The second main point of entry to the mine, the Mill level tunnel, is situated in San Miguel County adjacent to the Pandora mill, a short distance east of the town of Telluride, at an elevation of 9,060 feet. Virtually all of the production activities during the Idarado years occurred in the portion of the mine that is in San Miguel County. At its peak, the mine employed over 400 persons, with workers entering the mine via the Treasury Tunnel on the Red Mountain side of the divide and the Mill level tunnel on the Telluride side. Many of the miners commuted from towns over 70 miles away.

Authors' Note: *The authors of this article had the great fortune of working at the Idarado Mine early in their professional careers. This experience provided a solid background in mining and geology that has served them well through the years. It was a fascinating place to work: abundant with history and in a setting that surely rivals any other mining district in the world for its natural beauty. Because of the age and extent of its workings and the harsh environment, the mine presented special engineering challenges to modernization. From a geologic standpoint, though, it was rich in examples of ore deposition that were straight out of a textbook. But what they remember most are the crystals, and what an incredible profusion of crystals there was! They seemed to be everywhere, in almost every stope, raise, and drift. At times they were so plentiful that a person could, if he had the time, fill a powder box with crystals on each and every trip underground. After a while, however, one became numb to this extraordinary opportunity and would only pick up the occasional attractive specimen. Only years later did the authors realize how lucky they were to have experienced this special place.*

History

The Idarado Mine consisted of a collection of properties operated by a succession of companies from the late 1800s until 1953, when they were finally consolidated under one owner. The Black Bear vein, outcropping at an elevation of almost 12,000 feet, was discovered in 1894 and initially was an important producer of silver. As the mine went deeper, however, there was a change in mineralization whereby the silver content decreased and base metal content increased. The mine was operated sporadically thereafter but by 1934 had closed down and been abandoned (Ramsey, 1973). With a second world war looming, one group was convinced that there would be a growing demand for copper, lead, and zinc, and they reasoned that the down dip extension of the Black Bear vein represented an excellent base metal exploration opportunity. Leading the group was a well respected Newmont Mining Corporation geologist named Fred Searls.

Together they began consolidating the fractional land position in the area.

The Treasury Tunnel had been started in the early 1900s with the intent of intersecting the Black Bear and other veins at a lower elevation, but it was stopped more than 3,200 feet short of its target. In 1939 Newmont approached the Sunshine Mining Company, a major silver producer in northern Idaho, with a joint venture proposition to extend the Treasury Tunnel and explore the Black Bear vein at depth (Hardy, 1940). Sunshine accepted, and the Idarado Mining Company, a combination of "Idaho" and "Colorado," was born. The results of some initial core drilling in 1940 were discouraging, however, and after a two-year period of inactivity, Idarado leased the property in 1943 to the Metals Reserve Company, which had been established by the War Production Board to encourage output of metals needed for the war effort. The Treasury Tunnel was soon extended to 8,670 feet, where it intersected the Black Bear vein. Drifting on the vein exposed mostly low-grade ore, but plans were nevertheless made in 1944 to drive a raise from the Tunnel level up to the 6 level of the old Black Bear workings. By this time, the lease with Metals Reserve had been terminated, and Sunshine decided to sell its position (Hardy, 1945). Newmont went on to complete the Treasury raise, later known as "T-raise," in 1945. The raise was in ore all the way, and production plans were soon made that included the construction of a new mill on the Red Mountain side. By 1946, production had reached almost 66,000 tons per year and a small profit was realized. By 1949, production had doubled, and in 1950, Newmont sold a 20 percent interest in Idarado to the Homestake Mining Company (Ramsey, 1973).

On the other side of the divide, the famous Tomboy and the Smuggler Union mines had closed by 1928. In 1936 Veta Mines, Inc., acquired the Smuggler Union and in 1939

Upper left:
Calcite and pyrite, 7 cm wide
Idarado Mine, Ouray County
Benjy Kuehling collection, Jesse La Plante photo

Lower left:
Discrete particles of gold in quartz are typical for Idarado. This specimen is 5 cm wide and is from the AR 1042 vein.
John Trujillo collection, Jesse La Plante photo

Below:
Quartz variety amethyst, 6.6 cm tall
Collector's Edge, Inc. specimen, Jeff Scovil photo

Facing page:
Longitudinal section of the Argentine and Black Bear workings of the Idarado Mine
J. Walter Socha drawing

purchased the Tomboy (Raines, 2000). In 1940 Veta Mines was reorganized into Telluride Mines, Inc. By 1941, the company's mill at Pandora was producing more than 500 tons per day. Between 1945 and 48, Telluride Mines drove the Mill level tunnel. Completion of this tunnel allowed for more efficient development and mining of the lower levels of the mine, which by then was producing more revenue from lead and zinc than from precious metals (Hillebrand, 1957).

In 1953 Idarado acquired Telluride Mines, which effectively consolidated all of the best mining properties in the district. Management then formulated a plan to connect the Red Mountain and Telluride operations underground, shut down the mill on the Red Mountain side, and modernize the Pandora mill to serve the combined properties. By 1959 the project had been completed, and Idarado had become a much more efficient and lower-cost operation (Ramsey, 1973). The company operated at a profit throughout the 1960s and into the 70s, producing about 1,750 tons per day from a combination of vein and base metal replacement deposits.

In addition to the Black Bear, Tomboy, and Smuggler mines, other lesser-known properties were included in the consolidated Idarado holdings. One such mine was the Barstow, which was located on the Red Mountain side of the divide. During the 1970s, Idarado hauled a portion of the Barstow dump to its mill for processing, but serious metallurgical problems quickly put a halt to this activity. The Barstow's productive period pre-dated the formation of Idarado; some of the historic finds from the Barstow are discussed on page 58. Other small mines in the consolidated holdings that never made a comeback were the Liberty Bell, Handicap, Cimarron, and Ophir.

During its lifetime, the Idarado Mine was one of the leading producers of lead, copper, and zinc in the state of Colorado, and a sign on the old trestle above the entry to the main office complex on the Red Mountain side proudly proclaimed, "We mine the metals that make America strong." On July 18, 1978, Idarado Mining Company announced that

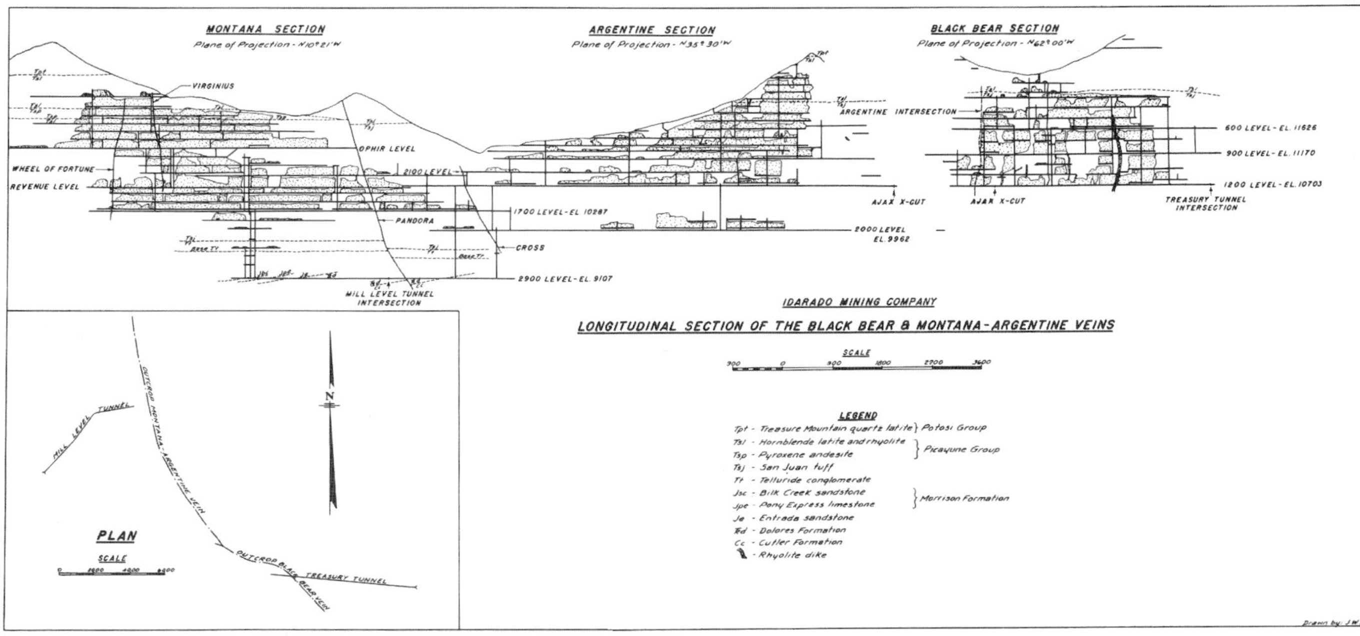

it would close the mine by the end of September. The announcement came as no surprise. As with many base metal mines in the west at that time, escalating mining costs, depressed metal prices, and higher smelter charges had led to its demise. The mine's closure was the knockout punch to a local mining industry that had seen the Camp Bird (see page 30) idled the year before, and the Sunnyside Mine near Silverton (see page 68) deluged by Lake Emma just a few months earlier. It brought an end to more than three decades of continuous mining and an associated period of unprecedented mineral specimen production.

At the time of its closure, Idarado Mining Company held a land position covering several thousand acres in portions of three counties in southwestern Colorado. In 1983, the State of Colorado filed a natural resource damage claim against Idarado to address the historic environmental impacts of mining on the area. The company and the state subsequently agreed on a comprehensive plan to improve the water quality in the local rivers and streams and to remediate certain mining disturbances, all the while preserving the historic mining character of the region. This effort was completed in 2000, following which Idarado began working with the Trust for Public Land for the purpose of returning a majority of its private land holdings back to the public domain. In 2006, the company received the Bureau of Land Management's National Director's Award as one of the finest examples of responsible mineral resource development in the country.

Ore Deposits

Most of the Idarado Mine's production came from two primary veins: the Black Bear and the Argentine-Montana, the Montana being the northwest extension of the Argentine vein. The Cross vein, thought to be a down-dip extension of the Tomboy vein, was also an important producer, especially during the mine's final decade of operation. Three others, the Basin, Ajax, and Legal Tender veins, provided the remainder of the mill feed. These veins are part of the radial fault system that extends to the northwest from the margin of the Silverton Caldera towards the Stony Mountain Stock. The Black Bear vein, which is closest to the caldera, has a strike of approximately N 60° W, while further to the northwest. the Argentine vein strikes from N 35° W to N 10° W (in the Montana workings). The Black Bear and Argentine veins dip from 70 to 80 degrees to the southwest. The Ajax and Legal Tender veins are situated to the southwest of and are roughly parallel to the Argentine. They also dip steeply to the southwest. The Cross and Basin veins are situated to the northeast of the Argentine and strike N 50° W and N 75° W, respectively. These veins also dip to the southwest but at a flatter angle, typically 55-65 degrees. The veins vary significantly in width from as narrow a few inches to more than 15 feet. Two of the veins (the Argentine and Ajax) are unique in that each formed in association with earlier fine-grained andesite dikes. These six veins, in addition to replacement deposits adjacent to the Argentine, Basin, and Cross veins, provided virtually all of the mineral specimens that are now attributed to the Idarado Mine.

Mineralization

The vein deposits in the Idarado Mine are classified as epithermal, based on fluid inclusion work on quartz from the Argentine vein. This work indicated temperatures of emplacement in a very narrow range (280 degrees to 290 degrees Celsius) with no evidence of fluid boiling (Nash, 1975). The veins display classic epithermal textures, including banding parallel to sub-parallel to the vein walls and brecciation within the veins, both of which are also indicative of multiple stages of vein emplacement. However, the most notable epithermal features are the abundant vugs, which provided critical open space environments for the slow, steady precipitation of minerals, resulting in spectacular crystal specimens.

While the mineralogy of the various veins is similar and is dominated by only a few species, the specimens found in each vein can vary considerably in character (color, transparency, size, etc.) and crystal habit. Quartz and calcite are the most common gangue minerals, followed by fluorite, rhodochrosite, pyroxmangite, dolomite, barite, and hematite.

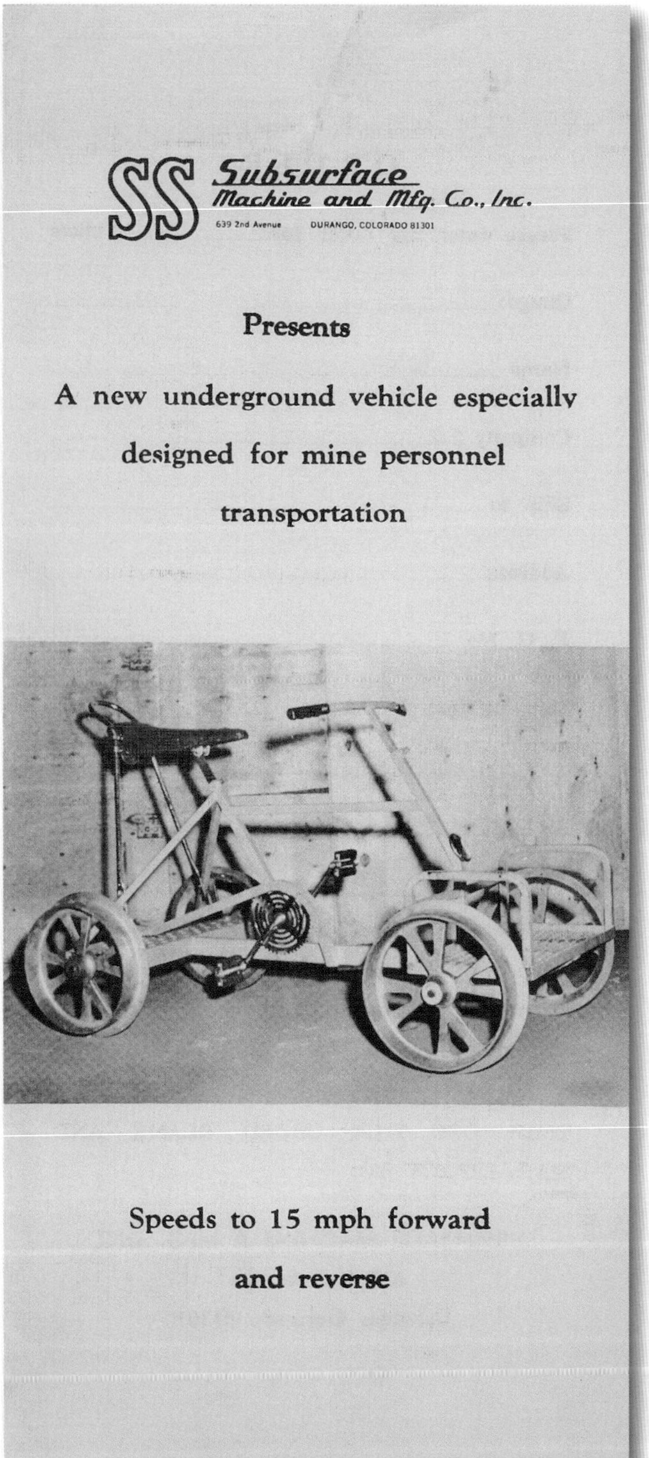

Minerals that are products of the hydrothermal alteration of volcanic wall rock fragments, gouge and "rock flour" seams within the vein include chlorite, sericite, and epidote. Major sulfide minerals include galena, sphalerite, chalcopyrite, and pyrite. Hübnerite, bismuthinite, and tetrahedrite are present only in minor amounts. Silver is present as a solid solution in argentiferous galena. Gold occurs as discrete particles in quartz and occasionally in sulfides.

Replacement type mineralization in the Idarado Mine occurs where certain veins pass from the sandstone of the Cutler Formation into the overlying Telluride Conglomerate. Replacement mineralization is typically at its widest and is best developed immediately above the base of the conglomerate. Less frequently, mineralization is also found in the underlying Cutler Formation, just below the contact. This style of mineralization has a manto-like appearance and may be present on both the hanging and footwall of the veins. The intensity of mineralization is dependent upon the chemistry of the vein and wall rocks, structural preparation, and distance from the vein. Near the vein, complete replacement of the cobbles and matrix is common, with the original conglomerate texture nearly obliterated. Further from the vein, replacement of only the most reactive matrix minerals typically occurs. At the intersection of major structures (such as the Basin and Cross veins), where there is an abundance of "plumbing," replacement mineralization is more concentrated and tends to be thicker and distributed over greater distances both vertically and horizontally. Typically, the grade of the replacement ore is directly proportionate to the grade present in the vein at that location. Because open spaces in the replacement mineralization are limited to the interiors of select cobbles, large crystals of any mineral species are rare. However, the replacement zones produced many excellent sphalerite, galena, and chalcopyrite crystals.

Mine Workings

While the locality given for most Idarado specimens is simply the mine name itself, a small number of specimens in collections identify the vein from which they came, and fewer still have the specific stope or other mine working noted. An explanation of the stope numbering system at the Idarado is provided here for those collectors with specimens attributed to specific locations in the mine. The system was based on the vein name, the level or sub-level that was directly below the stope and the stope block number. Each stope block was 250 feet in length along strike. Numbering increased from southeast to northwest, which was the general trend of the six vein systems. The beginning of a vein segment, where it could be determined, was assigned block number "01," as in the case of the Black Bear and the Argentine-Montana. For the other veins, the starting point for block numbering was more arbitrary. The Argentine-Montana vein system contained the most stope blocks, with over 2½ miles of almost continuous mining along its strike.

The level numbering system was based upon the approximate vertical distance, measured in feet, below the surface. The top-most level of the mine was the Black Bear 100 level, situated at an elevation of about 11,700 feet. For most of the mine's life, the 100 level adit was a natural ventilation exhaust. As a result of the cold, moist air rapidly expanding when it reached this level, perfect six-sided, paper-thin ice crystals to 15 centimeters across would form on the back and ribs along a short section of the adit and were a must-see stop for anyone on a mine tour. From the 100 level downward, level spacing was not uniform because of how the earlier mines had been independently developed, and level designations only approximated the actual distances separating them. The Treasury Tunnel level was designated the 1200 level, and thus was roughly 1,100 feet below the 100 level. The Mill level tunnel, the "bottom" of the mine, was designated the 2900 level. Other main levels (i.e. those equipped with rail haulageways) included the 300, 600, 800, 900, 1600, 1700, 2000 and 2400 levels. Because the levels

Right: Idarado's Pandora mill with the Meldrum Portal and dump in the background
Allan Young photo

Facing page: Circa 1970s advertising brochure for the type of subsurface mine bike used in the Idarado Mine
Subsurface Machine and Mfg. Co., Inc.

were so far apart, it was usually necessary to drive sub-levels between main levels so that stope heights could be minimized. Thus, other level designations such as 1400, 2200, 2600, 2700, etc. were also common.

Vein names in the stope numbering system were usually abbreviated, so the Black Bear became "BB," the Argentine, "AR," Cross vein, "CV," etc. For example, to designate a specific stope that was on the Cross vein, 200 feet below the 2400 level, in the 51 block, one would use "CV 2651."

There was no hoisting of ore or waste at the Idarado. A series of rock passes was used to drop ore to the 2900 level where it could be hauled to the mill. Waste was usually disposed of in empty stopes or was hauled outside where it was dumped. Major raises used for personnel and supplies included T-raise, which connects the 1200 level with the 900 and 600 levels; the Black Bear 600 raise, which connects the 600 level to the 300 and 100 levels; the Argentine 2413 raise, which connects the 2400, 2000 and 1200 levels; and the Argentine 2930 and 2030 raises, which connect the 2900 level with the 2000 level and the 2000 level with the 1200 level, respectively. These last two raises were the busiest during the 1970s, as a large share of the mining was being conducted from the 2000 and 2400 levels in the Cross and Basin veins. These two raises also provided a unique 5½-mile shortcut through the mine from Red Mountain to Pandora, instead of a 65-mile trip around the mountains by paved highway. This alternative route through the mine came in handy from time to time, such as when avalanches would close the Million Dollar Highway on the Red Mountain side. In this situation, mine and office workers, including accountants and secretaries and anyone else who wanted to go home that night, would then be loaded into "mantrip" cars and transported 3½ miles into the mine, down the two raises, and out another 2 miles to Telluride, where a bus would be waiting to take them home.

In addition to the Treasury Tunnel, the Mill level tunnel and the Black Bear 100 level, one other adit provided entry to the mine. The Meldrum Tunnel, whose portal is located west of the Pandora Mill and about 900 feet higher in elevation, crosscuts the Ajax vein at the 2000 level and intersects the Argentine vein at 2013 raise station. Originally begun as a narrow-gauge railroad tunnel through the divide, the Meldrum Tunnel was occasionally used to bring equipment and supplies in on the 2000 level and for disposal of waste rock. Because of the steady outflow of relatively clean mine water, beavers made several attempts to construct dams inside the tunnel, with predictably disastrous results (for the beavers).

The mine workings were spread out over a large area, and this created challenges to the supervisors, mechanics, engineers, and geologists who frequently had to visit many working places in a shift. The solution to this was the "mine bike." A quadricycle really, it was constructed of tube-steel, had flanged aluminum wheels and a pedal and sprocket arrangement driving the rear axle. The cycles were light enough to put onto a skip or pull off the track and onto a wide spot in the event a train was coming. There was a platform fixed over the rear axle in the back for a second person to ride standing up, and a basket affair on the front to haul tools, survey gear, and even powder boxes full of crystals. While providing an efficient means of travel around the mine, the bikes could also be dangerous. With attainable speeds in excess of 15 miles per hour, bike derailments were a common occurrence.

The veins mined at the Idarado were steeply dipping and relatively narrow (typically less than 10 feet) in width. Good vein continuity and competent walls presented ideal conditions for the application of a mining method known as "shrinkage stoping." This method involves advancing a stope upwards by drilling and blasting the ore above one's head in horizontal cuts, while leaving enough broken ore in the stope to stand on. Because broken ore takes up more volume than ore in place, the removal of a certain amount of excess volume or "swell" before each blast is necessary. Once a stope reaches its final height, the broken ore in the stope is drawn out from the bottom. This system, though fairly labor intensive, worked well in the veins, making the Idarado one of the best examples of the use of this method in the world.

Fortunately for collectors, this mining method also lent itself well to specimen recovery. A fresh face of unbroken ore extending the entire length of the stope was exposed after each cut, providing an excellent opportunity to examine the vein for vugs and remove any crystals one might find.

During the mid-1970s, Idarado incorporated a different mining method that was designed to mine a large, flat-lying, irregular-shaped zone of base metal replacement ore known as the BA 2652 replacement orebody. This orebody was situated at the base of the Telluride Conglomerate, adjacent to the intersections of the Cross and Basin veins below the 2400 level. The method was referred to as "room-and-pillar" and involved driving horizontal openings through the ore utilizing rubber-tired equipment, leaving pillars to support the back where required or where the grade of the rock was too low to mine. The method was more damaging to crystal specimens because of the larger amount of explosives used per volume of rock, but the face could still be inspected after each blast as was the case with vein mining. "Backstoping" and "benching" were the methods used to mine a replacement orebody adjacent to the Argentine vein, known as AR 2515.

Specimen Production

Fortunately for mineral collectors, Idarado management took a fairly liberal position on mineral collecting by mine workers. While high-grading gold was obviously prohibited, the company for many years allowed employees to collect non-metallic minerals specimens for their own collections as long as it did not interfere with their duties. It was a nice perk for those working underground. Most of the development and stope miners were paid on an incentive bonus system based upon the tonnage they mined or the footage of drift or raise they drove. The best of these miners were thus highly motivated and often did not stop for lunch, let alone take the time to collect crystals. Nor did many of the miners typically carry tools other than a pipe wrench, so they were ill-equipped to extract crystals from the hard quartz veins. Nevertheless, many excellent crystal specimens did find their way into lunch buckets and onto shelves at home. While almost everyone who collected crystals in the mine eventually sold a few specimens, selling mineral specimens evolved into a lucrative side business for some. The challenge was to get large quantities of crystals delivered to the surface without others knowing about it. One clever individual, knowing that the supply crew would collect and haul out the waste from underground toilets in appropriately labeled powder boxes, started shipping boxes of crystals marked in the same manner. This system worked well for awhile, as no one dared to investigate these boxes. Curious as to why someone would perform one of the most unpleasant jobs in the mine for him, one of the supply men eventually exposed the scheme.

A few workings in the mine achieved almost legendary status for their quantity and quality of crystals. One such place was the Argentine 2714 stope. The dike adjacent to the vein in this stope carried an open vug that was remarkably continuous both in strike and dip. The vug was filled with long, slender, milky to transparent quartz crystals to several centimeters in length that made very attractive specimens.

Another prolific but lesser known specimen producer was the Black Bear 1408 stope. This stope was mined out in 1968, leaving a large void where the vein had been. Years later, however, it was still possible for one to negotiate his way down through the debris in the raise to an opening into the old stope and then crawl across the raise pillar into the remains of a large vug, all the while being careful not to fall into the 200-foot chasm below. Referred to as the "secret vug," it was a dangerous place, and only a few individuals knew about it. One mine geologist who often visited the vug left its actual location with his wife in the event that, if he ever failed to return home, she could tell the rescuers where to look for him. In the early 1970s, several amethystine quartz crystals were removed from this stope, with one spectacular specimen that reached 48 centimeters in length.

It is difficult to estimate the quantity of mineral specimens that were saved during the life of the Idarado Mine. Production likely exceeded 100 tons of crystals, perhaps 90 percent of which were quartz specimens. In general, most of the quartz, calcite, fluorite, and gold specimens came from the veins, and the best sulfide crystal specimens came from replacement deposits adjacent to the veins. While the mine was operating, Idarado specimens were plentiful and inexpensive. As a youngster, one of the authors (JT) was among the many children doing a brisk business selling minerals to tourists along the Million Dollar Highway. However, with the mine having closed more than 30 years ago, good specimens have become difficult to obtain.

The Minerals

While more than 35 valid mineral species have been reported from the Idarado Mine, the following discussion includes only the more common, collectable species for which the mine is best known.

Quartz

Quartz was ubiquitous and by far the most collected mineral species at the Idarado. Quartz from the Black Bear vein often occurred in large plates of milky, thick, lustrous crystals with complex prism faces curving toward the termination. In the BB 1408 and other Black Bear stopes, translucent to transparent, tapered amethyst crystals exceeding 10 centimeters in length were collected. Many of these are doubly-terminated. In the BB 1616 stope, quartz crystals to 20 centimeters in length are associated with lustrous crystals of chalcopyrite and sphalerite. In the same stope, large milky quartz crystals at one time lined a vug large enough to stand in that continued upward as far as one could see. In the BB 322 stope, water-clear, doubly-terminated crystals to 10 centimeters in length with fluid inclusions were found during the early 1970s. Quartz groups containing long, slender, amethyst crystals associated with short, thick milky crystals with amethyst tips were also recovered from the Black Bear vein.

The Argentine vein produced quartz crystals similar to those found in the Black Bear. Large plates containing short, thick milky crystals were found in a number of stopes. Many specimens contain two or more generations of quartz, with larger crystals covered by dull or sugary coatings of later quartz. In some cases, the later generation of quartz only covers the prism faces of the larger crystals. The Argentine vein produced groups and intergrowths of reverse sceptered and doubly-terminated, translucent to transparent crystals as well as groups of crystals to 5.5 centimeters in length with sceptered and reverse-sceptered amethyst tips. The Idarado Mine also produced Japan-law twins to 9 millimeters across on calcite that likely originated in the Argentine vein.

The AR 2714 stope produced a large number of plates with long, slender, milky to transparent quartz crystals to 15 centimeters in length. On the 800 level of the Argentine, between the 38 and 42 blocks, milky quartz crystals 10 centimeters long in plates measuring 49 centimeters across were collected in 1978 (Murphy, 1979). Small jackstraw groups of sparkling, transparent, doubly-terminated quartz crystals associated with sulfides, epidote, and chlorite were frequently found in vugs randomly distributed in the replacement ore adjacent to the Argentine vein.

A unique occurrence of quartz epimorphs after an unidentified mineral, possibly anhydrite, was found in the AR 1408 stope during a period of a few weeks in the spring of 1975. Many of these were single, elongated, hollow casts of drusy quartz that resembled a knife scabbard, while others were composed of a jumble of angular casts to 30 centimeters across that the miners referred to as "angle iron quartz." One of the authors (AY) recalls more than 100 of these specimens lying on the muck pile after one blast. Quartz was also found as crystalline coatings on other minerals such as calcite, gypsum, barite, and most notably, cubic and octahedral fluorite. In many cases, the underlying mineral was dissolved away, leaving empty shells or casts in the shape of the original crystal. In some instances, remnants of the original mineral remain inside the casts. One specimen of quartz after anhydrite from the Argentine vein resembles an open book, with the "pages" measuring 10 by 6 centimeters.

Quartz crystals containing phantoms were occasionally

Above: Quartz, field of view 6 cm tall
Dave Bunk specimen (ex Gary McWilliams), Jesse La Plante photo

Upper right: Quartz, 16.5 cm tall
Collector's Edge, Inc. specimen, Jeff Scovil photo

Lower right: Quartz epimorph after anhydrite, 11.2 cm tall
AR 1408 stope
Collector's Edge, Inc. specimen, Jeff Scovil photo

collected from the Black Bear and Argentine veins. One plate consists of more than one hundred translucent to transparent crystals to 1.5 centimeters long, each exhibiting a phantom whose six pyramidal faces are highlighted by an underlying layer of an unidentified brown mineral.

The Cross vein did not produce a large number of quartz specimens, nor did the replacement orebody that was adjacent to the intersections of the Cross and Basin veins. However, the Basin vein itself did yield many fine quartz groups. Large plates of milky to water-clear quartz crystals to 7.5 centimeters in length were found near the top of BA 2952 raise. Many of these plates appear to be composed of doubly terminated quartz crystals packed together in parallel, with the back side of the plate also covered with terminations. Quartz from the Basin vein is commonly associated with microscopic crystals of sphalerite, galena, and chalcopyrite. During the driving of a drift on the 2900 level in the Basin vein in 1970, one of the authors (AY) recalls standing on top of a locomotive with the upper part of his body completely engulfed by a large vug that was in the back. With his mine light reflecting off thousands of transparent quartz crystals, it was like being in the center of a natural crystal chandelier.

In the Ajax vein, stopes above the 2000 level (specifically the AJ 2008 and AJ 2009) were known for large quartz vugs lined with thick, milky quartz crystals to 7 centimeters in length. Small calcite scalenohedra typically cover much of the quartz, and both in turn are dusted with microscopic pyrite crystals. These specimens are easily recognizable and can be confidently attributed to this locality. The Legal Tender vein, located along the same trend but southeast of the Ajax, did not produce a significant number of collectable minerals.

Calcite

Good calcite specimens from the Black Bear vein were not common although plates 20 centimeters across reportedly host pink rhombohedra exhibiting split crystal growth, thereby imparting a jagged appearance to the crystals (Raines, 2000).

The Argentine vein yielded excellent specimens of "manganoan" calcite. The AR 2724 stope reportedly produced crystal groups to 15 centimeters across (Raines, 2000). During the mid-1970s, the AR 2714 stope produced a number of matrix specimens containing groups of pink scalenohedra. Parallel growth features on the scalenohedral faces result in multiple terminations that lend a jagged appearance to some of the crystals. The calcite is commonly associated with snow-white, thinly bladed groups of barite crystals to about 2.5 centimeters on edge and transparent quartz crystals, many doubly-terminated, on a matrix of fine-grained chlorite and epidote. These specimens were recovered from a series of vugs along the contact of the vein and accompanying dike. One group featuring a 15 centimeter crystal is certainly among the finest examples of manganoan calcite from any locality.

Calcite was the predominant gangue mineral in the stopes of the Cross vein. A number of specimens reported from the CV 2958 stope consist of chains of complex, stacked scalenohedra to 7 centimeters in length. The crystals are typically translucent to transparent near their edges, with an opaque white band running through the center of the chain. Specimens measuring over 21 centimeters in length were collected. Typically covering the surface of these specimens is a light dusting of microscopic pyrite crystals. Cross vein stopes below the 2400 level also produced specimens composed of complex intergrowths of flattened, twisted, and bent scalenohedra reaching lengths of 14 centimeters. These calcite specimens are translucent and typically light-pink in color, and some are heavily coated with iron oxides.

The Cross and Basin veins produced a large number of

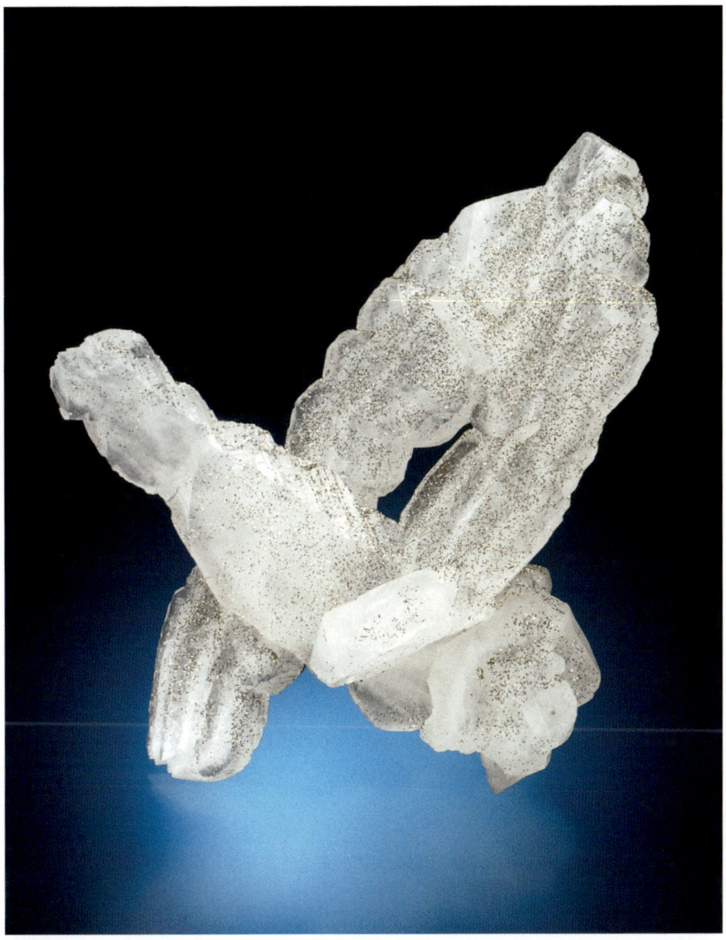

Upper left:
Pyrite on calcite
5.4 cm tall
Ajax vein
Benjy Kuehling collection
Jeff Scovil photo

Upper right:
Calcite with pyrite
9 cm tall
Cross vein, Idarado Mine
Dave Bunk collection
Jeff Scovil photo

Right:
"Manganoan" calcite
Main crystal 15 cm tall
AR 2714 stope
Collected by Jack Crawford
Dave Bunk collection
Jesse La Plante photo

Facing page, left:
Quartz, 12 cm tall
Collector's Edge, Inc.
specimen
Jeff Scovil photo

Facing page, right:
Quartz after fluorite
9.5 cm tall
Black Bear vein
Dave Bunk collection
Jesse La Plante photo

Left: Quartz on chalcopyrite, 6.5 cm tall
Black Bear vein
Dave Roter collection, Jesse La Plante photo

Above: Idarado engineer Andy Hoover (grandson of President Herbert Hoover) In a Cross vein shrinkage stope, ca. 1970
Allan Young photo

Below: Sphalerite with chalcopyrite, 4.5 cm wide
collected by Allan Young in the AR 2515 replacement orebody
Benjy Kuehling collection, Jesse La Plante photo

Facing page: Galena with sphalerite, 5 cm tall
Basin vein 2652 replacement orebody
John Trujillo specimen, Allan Young photo

specimens of manganoan calcite from stopes below the 2400 level during the 1960s and 70s. Plates of scalenohedra to 60 centimeters across were recovered, with individual crystals measuring up to 6 centimeters in length. These crystals typically exhibit a blunt, rhombohedral termination. Manganoan calcite specimens from the Cross and Basin veins are easily distinguished from those from the Argentine by their distinctive crystal habit and peach-pink color. These specimens are by far the most common and recognizable of the Idarado manganoan calcites.

Plates of complex scalenohedral calcite crystals covered by a thin layer of manganese or iron oxides were recovered from the CV 2655 and 2656 stopes during the 1970s. These were found in clay-filled vugs several yards long. While interesting, the plates were difficult to recover intact and were not usually attractive; so large numbers were not collected.

Light-pink, scalenohedral casts composed of manganoan calcite were found in the BA2652 replacement orebody, together with unusual parallel growths of small, acicular crystals of manganoan calcite on calcite scalenohedra.

The Ajax vein produced doubly terminated, scalenohedral calcite crystals to 7 centimeters in length that are colorless and transparent to off-white and opaque. The crystals are commonly attached to quartz plates and many are twinned on {0001}. Many of these single crystals and crystal groups are partially or completely covered with a thin crust of tiny, lustrous pyrite crystals (Raines, 2000). Some marcasite may also be associated with these groups.

Gold

Mayor (1978) estimated that more than 735,000 troy ounces of gold were produced from the Idarado Mine from

1946 through 1976, most of this came from the Argentine vein. Likewise, most of the specimen gold also came from the Argentine, with lesser amounts from the Black Bear. Virtually all of the gold was found above the 1200 level. Specimen gold has not been reported from the Cross, Basin, and Ajax veins or from any of the replacement orebodies, most of which were situated well below the 1200 level. Nor has gold been reported from the limited workings on the Legal Tender vein.

Gold from the Black Bear vein reportedly occurred primarily in post-ore fractures and so was deposited late in the paragenetic sequence. Gold occurred in small quartz stringers cutting the footwall of the vein, in greenish-colored quartz in altered wall rock, in clay slips as thin gold wires, and in vuggy white quartz as mossy aggregates of gold with calcite and fine-grained galena (Eckel, 1997). In the BB 319 stope, gold was occasionally found as masses of wires to a few centimeters. The Idarado Mine has also produced attractive gold specimens that exhibit a boxwork of bright-yellow gold leaf in small veinlets of quartz and calcite that likely originated from the Black Bear vein.

The AR 1042 stope was a prolific producer of specimen gold during the mid 1970s. The raise that serviced this stope was driven from the 1000 level to the surface, and many of the raise cutouts contained visible gold. The AR1042 slusher drift above the 1000 level produced coarse gold from a location 125 feet from the chute. Here, the gold occurred as irregular masses filling void spaces between quartz crystals, in some cases showing perfect casts of the crystal faces. Ribbon gold filling small vugs in quartz were also found in the AR 1042 stope. Some of the gold from this stope has a greenish or reddish tarnish.

Other prolific gold producers included the AR 1124, 1125 and 1126 stopes. Specimens from these stopes were frequently cut into slabs and used in jewelry, and some show cockscomb quartz veining with gold filling the voids, thereby highlighting the jagged outline of the crystals. Inclusions of gold in transparent quartz crystals were also reported from the Argentine vein, as was wire and sponge gold. However, most of the gold produced from the Argentine vein consists of irregular blebs and streaks of the yellow metal in massive vein quartz, typically associated with fine-grained sulfides, rhodochrosite, and pyroxmangite.

Sphalerite

Sphalerite was the most common sulfide mineral in the veins of the Idarado Mine. Olive-green sphalerite was observed in a drift heading in the Basin vein on the 2000 level. Greenish-black to rosin-colored crystals of sphalerite exceeding 2.5 centimeters across were collected in the Argentine vein and replacement orebodies below 2400 level, as well as in the BA2652 replacement orebody. These crystals occur in parallel growths several centimeters across or as epitactic overgrowths on chalcopyrite. Crystals are often complex and exhibit numerous forms. Other associated minerals include galena, quartz calcite, epidote, and chlorite (Raines, 2000). These specimens are virtually impossible to distinguish from those of the nearby Camp Bird Mine.

Chalcopyrite

Chalcopyrite was the main copper-bearing mineral at the Idarado Mine. In the Black Bear vein, complex intergrowths of crystals were found that have a bluish gray to almost black tarnish or thin coating of some as yet unidentified mineral, possibly chalcocite (Raines, 2000) or digenite. On some specimens, the chalcopyrite is lustrous and untarnished and is associated with transparent quartz crystals.

Sharp, striated tetrahedrons to 5 centimeters across were found in the Argentine vein replacement stopes below the 2400 level. Many of these have a similar tarnish or coating to those found in the Black Bear. These tarnished crystals are frequently associated with small needles of yellow-green epidote (Raines, 2000). In the AR 2515 replacement orebody, bright, euhedral crystals to 1.2 centimeters were also found, as were etched crystals up to 4 centimeters across.

Chalcopyrite was also found in the BA 2652 replacement orebody. One specimen consists of several bright,

interlocking chalcopyrite crystals to 2 centimeters intergrown with transparent quartz crystals. Small crystals of galena and sphalerite are scattered about the chalcopyrite crystals.

Galena

Galena was an important ore mineral in the veins and replacement orebodies of the Idarado Mine and carried most of the silver that was in the ore (1.8 to 3.0 ounces per ton). In massive form, specimens typically exhibited warped cleavage planes. Galena was not as common in crystals as sphalerite or chalcopyrite. Specimens from the replacement orebodies typically exhibit some degree of hoppered or skeletal growth and are usually associated with chlorite group minerals. Many crystals are twinned, and most have a brilliant luster. One 3-centimeter specimen from AR 2515 replacement orebody exhibits twinned octahedral and cubic faces with internal skeletal growth features.

Fluorite

Massive, pale-green fluorite was common in the Cross and Basin veins, although good crystals were rare. Those rarities are typically 1 to 2.5 centimeter octahedra, some of which are modified by trisoctahedra at the points at which the octahedral forms intersect. The faces of the crystals are often skeletal, lightly etched, or both. Fluorite occurring in pale purple cubes 0.5 to 1.5 centimeter across on sugary quartz matrix was also reported (Raines, 2000).

Fluorite occurred sparingly in the BA 2652 replacement orebody as transparent, colorless cubes to 1 centimeter across with trapezohedral modifications on the corners. In the AR 2515 replacement orebody, light blue-gray, translucent cubes to 2 centimeters on edge on small white dolomite crystals were recovered during the last year of the mine's operation.

Fluorite and Quartz from the Barstow Mine, Ouray County, Colorado

by Horace B. Patton

excerpted from the *Quarterly of the Colorado School of Mines*, October 1907, Vol. 2, No. 2

In the summer of 1906 a large number of specimens composed of fluorite and quartz were obtained from the Barstow mine, Ouray County, Colorado, through the kindness of Mr. John Geisel, the superintendent, who not only gave access to the mine, but freely furnished men, tools and powder and considerably laid aside and preserved for shipment many good specimens as they were disclosed in the development of the mine.

The fluorite occurs in rough octahedral crystals that vary in size up to one and one-half inches, averaging about one inch. The roughness of the crystals is due to the fact that the apparently octahedral form is not really such but the crystal is built up in pyramid-fashion out of miniature crystals each of which consists of a cube with its edges beveled by the faces of a tetrahexahedron. These miniature cubes measure from one to two millimeters in diameters. These fluorite crystals have a very delicate color that varies from pale green to a fairly deep green. Owing to the rough surface they are translucent rather than transparent.

The quartz of these specimens is pseudomorphous in character. The original mineral had the shape of flat rhomboidal plates very suggestive of the habit frequently assumed by barite, in which the basil pinicoid occurs in combination with prism of about 101 degrees. While the writer is of the opinion that this is really a case of pseudomorphism after barite, satisfactory proof thereof is hardly possible. The original tablets must have been very thin, not more, perhaps, than one millimeters; but the original thickness has been about doubled by the deposition of minute drusy quartz crystals over the surface. The interior is often characterized by a gray color in marked contrast to the outer drusy crust which is snow-white. These drusy quartz pseudomorphs attain a maximum size of about one to two inches. The plates arrange themselves in sometimes more or less parallel bunches, but more usually present loose reticulated masses.

Fluorite, 7 cm wide
Barstow Mine, part of the modern Idarado Mine complex, San Miguel County
Colorado School of Mines Geology Museum collection, Jesse La Plante photo

To a slight extent calcite has formed in connection with these aggregates of fluorite and quartz. It is whitish and translucent. The form has the appearance of that of a slender scalenohedron. In reality these likewise are built up out of small individual crystals that grow together as not quite parallel aggregates. The actual forms are rather obscure, but a steep scalenohedron is prominent. These roughly scalenohedral, complex individuals measure at times from one to three inches in length.

The order of deposition of these minerals is as follows: First, pseudomorhic quartz, presumably replacing barite; second, fluorite third, calcite. In addition to these three may be mentioned a little quartz mingled with calcite that occurs like a fine powder sprinkled over the surface of some of the fluorite masses.

The exquisite contrast in color between the delicate green fluorite and the snow-white quartz, together with the delicacy of form and setting, make these specimens unusually beautiful. They are, in fact, among the most attractive in the School of Mines mineral cabinet. ...

The Barstow mine is a gold-silver property in which the proportion by weight of gold to silver is about one to four. The ore occurs in a quartz vein that cuts an andesite breccia and consists of pyrite, a little chalcopyrite and small but greatly varying amounts of sphalerite. The gold is partly free milling and partly in the form of a black telluride that resists amalgamation. The portions of the vein that contain much fluorite are practically barren of gold and silver.

Above: Epidote and quartz, 6 cm wide
Dave Bunk collection, Jesse La Plante photo

Right: Rhodochrosite, 4 cm tall
1126 stope, Argentine vein
John Trujillo specimen, Allan Young photo

Pyrite

The veins of the Idarado were relatively iron-poor in comparison to other deposits in the area, so pyrite crystal specimens were not common in the mine. Specimens consisting of microscopic crystals of pyrite encrusting both simple and complex calcite scalenohedra were recovered from the Argentine vein. Irregular patches of dark, crystalline pyrite were occasionally found on plates of milky quartz crystals typical of those from the Argentine vein.

Raines (2000) reported that the Ajax vein produced several habits of pyrite, including groups of curved cubes arranged in a stair-step pattern, 1 to 2 centimeters across, on calcite. The Idarado also produced lustrous pyritohedra to 2.5 centimeters associated with transparent quartz crystals as well as pyritohedrons to 3 millimeters on calcite scalenohedra.

Barite

Attractive, snow-white, thinly bladed groups of barite crystals to 2.5 centimeters on edge were associated with quartz and manganoan calcite from the AR 2714 stope.

Rhodochrosite

While veinlets of massive rhodochrosite were common in some stopes in the Black Bear and Argentine veins, good crystals were rare. Curved, light-pink rhombohedra to 3 millimeters were occasionally found on transparent quartz crystals in the Argentine vein. In the AR1042 stope, crystalline masses of light-pink, flattened rhombohedra to 8 millimeters were recovered. In the AR1126 stope, individual rhombohedra to 4 millimeters were associated with calcite and galena.

Other Minerals

Epidote was pervasive throughout the mine as a propylitic alteration of the wall rocks near the veins. It was common in the Argentine vein and replacement orebodies as well as the BA 2652 replacement orebody and was typically associated with transparent quartz, chalcopyrite, and sphalerite. Crystals were usually transparent but typically only a few millimeters in length. Pale, brownish-red crystals from the 2400 level of the Argentine vein, previously thought to be piemontite, were later determined to be manganese-rich epidote (Eckel, 1997).

During the mid-1970s, hübnerite blades to 2 centimeters were found in massive quartz near the intersection of the Montana and Virginius veins on the 1700 level.

Pyroxmangite, erroneously referred to by the miners as "rhodonite," was a common constituent of the veins in the Black Bear and Argentine stopes. It also occurred in the replacement orebodies as nodules to 10 centimeters across. An apparently singular find of small, tabular crystals of pink dolomite to 1 centimeter across was collected from the Argentine vein during the 1970s. Acicular crystals of bismuthinite 1 to 2 millimeters in length were reported from the Basin vein below the 2400 level. Massive, specular hematite was common in several of the Cross and Basin vein stopes.

Conclusions

The Idarado Mine was one of the best known and most prolific specimen producing mines in the western United States, with many tons of quartz, calcite, and sulfide crystals finding their way into public and private collections worldwide. Since its closure more than 30 years ago, quality specimens that were once relatively common and inexpensive are now becoming harder to obtain on the mineral market. While precious and base metal resources do remain in the mine, its current reclaimed and protected status probably preclude it from ever being reopened.

ACKNOWLEDGMENTS: *The authors wish to thank Dave Bunk, Bryan Lees, and Benjy Kuehling for allowing them access to study and photograph their Idarado collections. Thanks also to Tom Rosemeyer for reviewing the article.*

Rhodochrosite from the San Juans

by Barbara L. Muntyan

Rhodochrosite is perhaps the most sought-after mineral species from the San Juan Mountains of southwestern Colorado, both for private mineral collectors and museums. Occurring in well-crystallized examples, the mineral's color is most often pale pink, edging toward strawberry-red in a few specimens. Rhodochrosite crystals are typically found as sharp rhombohedra and are usually implanted on drusy white quartz. San Juan rhodochrosite, especially from vugs in the Sunnyside Mine north of Silverton, is sometimes found associated with pale green fluorite octahedra. Rhodochrosite also occurs in association with sphalerite, pyrite, chalcopyrite, calcite crystals or crusts, and even (rarely) gold. Very often, especially on specimens from the Sunnyside and Grizzly Bear mines, rhodochrosite crystals are completely coated with drusy quartz.

The Sunnyside Mine was certainly the San Juan's most prolific producer of rhodochrosite, which occurred in many zones in the expansive workings. But other, smaller mines also yielded attractive and well-developed crystals of the pink mineral. The Grizzly Bear, Mountain Monarch/ Mickey Breen, and Daniel Bonanza mines, all in Ouray County, and the Brooklyn Mine in San Juan County are among the lesser known rhodochrosite localities, but interestingly with a few exceptions, both the Camp Bird and the Idarado mines are lacking in collectible rhodochrosite specimens. Both mines occasionally produced specimens of pale pink crusts on quartz, but there are no published analyses of the pink material, which could fall in the rhodochrosite ($MnCO_3$)-calcite ($CaCO_3$) solid solution series.

Nearly all of the rhodochrosite from the San Juan's deposits is pale to medium pink, although a few crystals from the Belle Creole vein at the Sunnyside Mine, recovered in the 1970s, are cherry-red. The pale pink San Juan rhodochrosite formed at a relatively low temperature in the Silverton Caldera and contains varying proportions of manganese, calcite, and magnesium.

Sunnyside Mine, San Juan County

Rhodochrosite from the Sunnyside Mine (see page 68) was plentiful, commonly forming large, varied, and well-crystallized specimens. The original claims that were the foundation of the Sunnyside Mine were staked in 1873 on the sunny side of Hurricane Peak. At its peak, the mine workings stretched in a northwest-southeast trend more than 4 miles on six levels extending along the vein. The mine was the largest gold mine in the United States at the end of the nineteenth century and was a major producer of silver, lead, zinc, and copper.

In the 1960s the American Tunnel was driven to intersect the shafts of the older mine workings. During the

Upper left: Rhodochrosite on quartz, 3 cm tall
Grizzly Bear Mine, Ouray County
Dave Bunk collection, Jesse La Plante photo

Upper right: Rhodochrosite on quartz, 3 cm tall
Mountain Monarch Mine
Tom Hughes collection, Jesse La Plante photo

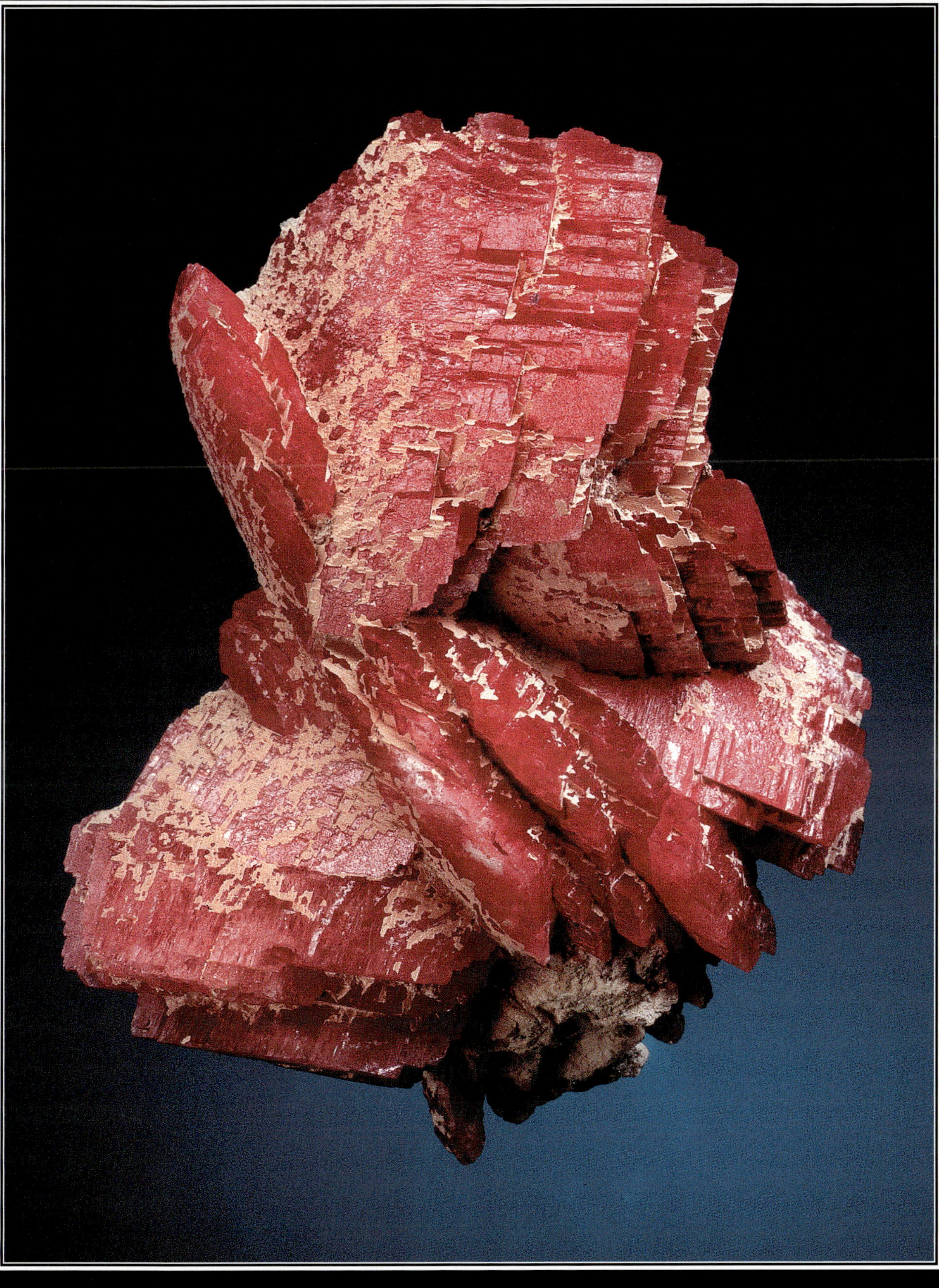

"Curved" Rhodochrosite Rhombohedra
8 cm tall
Sunnyside Mine, near Gladstone, San Juan County
Benjy Kuehling collection, Jeff Scovil photo

Upper left:
Rhodochrosite on gypsum, 6.9 cm tall
Sunnyside Mine, near Gladstone, San Juan County
Collector's Edge, Inc. specimen, Jeff Scovil photo

Above:
Rhodochrosite on quartz, 7 cm tall
Sunnyside Mine, near Gladstone, San Juan County
Marty Zinn collection, Jeff Scovil photo

Lower left:
Rhodochrosite on fluorite, 5 cm tall
Sunnyside Mine, near Gladstone, San Juan County
Private collection, Jesse La Plante photo

Facing page:
Vintage photo of the trestle at the Grizzly Bear Mine
Barbara L. Muntyan collection

expansion, miners encountered many rhodochrosite- and/or fluorite-bearing vugs in the upper workings of the mine, but good rhodochrosite crystals were not collected in the tunnel itself; thus, specimens stated as originating from the "American Tunnel" are more properly labeled, "Sunnyside Mine, near Gladstone, San Juan County, Colorado."

The finest specimens of Sunnyside rhodochrosite are sharp, light pink to strawberry-pink rhombs up to 5 centimeters on edge. They occur as single crystals or as intergrown clusters on plates of drusy quartz crystals. Nearly every major museum in the world boasts at least one such specimen, as do many private collections.

A rhodochrosite habit that is nearly unique to this mine is one of curved rhombohedra, which can form single crystals to 5 centimeters on edge or occasionally clusters. The curvature is pronounced and is similar to that of saddle-shape dolomite. Crusts of delicate, pale pink curved rhombohedra have also been found coating black sphalerite in Gilman, Colorado, and recent finds of rhodochrosite from the Wuton (Wudong) Mine in China's Guangxi Zhuang Region display a similar curvature.

Less common but also highly prized are clear gypsum (variety selenite) crystals that have rosettes of pink rhodochrosite embedded in them. The gypsum is so clear

that the rhodochrosite is plainly visible though both sides of the crystal. At least two pockets containing these specimens were found, and a total of perhaps ten such crystals were recovered.

Combinations of green octahedral fluorite crystals associated with rhodochrosite are also coveted by collectors. Generally the rhodochrosite rhombohedra in these combinations are light pink, and the fluorite octahedra are medium to emerald-green. In many of these specimens, the fluorite octahedra are the larger of the two species and are sprinkled with small rosettes of rhodochrosite. Groups up to 5 centimeters are not uncommon with the largest reaching 10 centimeters.

Various generations of Sunnyside rhodochrosite are commonly found between different generations of quartz and fluorite. These multi-generation combination pieces highlight the complex stages of mineral deposition in the veins of the Sunnyside Mine.

The Sunnyside Mine reopened three years after the catastrophic 1978 flood (see page 81). The top three levels of the mine had been abandoned, and mining concentrated on the D-E-F levels. Nearly all of the quartz at the Sunnyside occurs in druses or clusters of "points," but the F level of the mine produced atypical long, white quartz crystals reaching 5 centimeters in length with clusters of small, light pink rhodochrosite crystals on their prisms.

Sunnyside rhodochrosite is also found in association with sphalerite and other sulfides; in fact, the Sunnyside Mine produced literally thousands of rhodochrosite specimens, many in combination with other species from the mine.

Grizzly Bear Mine, Ouray County

The Grizzly Bear is an old mine; the original claims were located in 1875, some two years before Ouray County was carved out of the San Juan Territory. The claims are about halfway up Bear Creek from the Uncompraghre gorge, on the southwest slope of Engineer Mountain.

Mine development was hampered by the lack of a real road; the Grizzly Bear was accessed only by a long and difficult pack trail up the Bear Creek drainage, beginning at the Silverton-Ouray toll road. In spite of the transportation challenges, the mine produced approximately one million dollars in gold and silver ore between 1875 and 1910. Nonetheless, a series of attempts failed to obtain major financing for mine development, a better road, or a mill site.

As early as 1910, fine rhodochrosite crystals were reported from the mine. In one of the mine promotional reports, mining engineer Chester Ingersoll described, "… quartz crystals and very beautiful crystals of rhodochrosite (manganese carbonate)." Because mine promoters are not generally given to singing the praises of non-ore material, the rhodochrosite specimens found in that early period must have been truly impressive.

Over the years many prominent and ambitious Ouray businessmen were associated with the Grizzly Bear. George and Edward Wright, who built the opera house in Ouray with part of the fortune they had realized by discovering and developing the Wheel of Fortune Mine, owned the Grizzly from 1894 to 1896. They might have operated it for much longer had Edward not contracted pneumonia and died while working at the mine. George then sold the property to George Hurlburt, then president of the Bank of Ouray and part owner of the Bachelor Mine north of town.

Lars Pilkaer, a road and bridge builder and occasional miner, acquired the Grizzly Bear after it went into receivership just after World War I. Lars had worked at the Grizzly for the Wright brothers. He sold part of his interest in the mine to recently arrived Italian immigrant John Zanett (nee Giovanni Zanetti). Zanett, blessed with a silver tongue

Ouray County Rhodochrosite

Upper left:
Rhodochrosite on quartz
6 cm wide
Zanett Tunnel, Grizzly Bear Mine
Tom Hughes collection
Jesse La Plante photo

Lower left:
Rhodochrosite on quartz
5.6 cm wide
Grizzly Bear Mine
Dave Bunk collection
Jesse La Plante photo

Facing page, left:
Rhodochrosite, 2.8 cm tall
Mickey Breen Mine
Dave Bunk collection
Jeff Scovil photo

Facing page, right:
Rhodochrosite on quartz
6.25 cm wide
Mountain Monarch Mine
Barbara L. Muntyan collection
Jesse La Plante photo

and an expansive personality, worked for the next half century to obtain investment monies to develop the Grizzly Bear, and he did meet with occasional success.

Finally in the 1970s, John's youngest son Fred, who had made his fortune in the Texas oil patch, decided to buy all of the patented claims on the north side of Bear Creek. His plan was to build a haulage tunnel through the mountain and under Bear Creek (a distance of about 2 miles) that would intersect the old workings and solve the transportation problem once and for all.

It took them more than a dozen years to acquire the claims and commence work, but Fred Zanett and his partner Casewell Silver, a former United States Geological Survey geologist, managed to acquire 98 percent of the Grizzly Bear claims and drove the John Zanett Tunnel all the way to the old workings on the Grizzly Bear vein. The tunnel cut through various other veins and produced a lot of fine rhodochrosite specimens in the process.

Construction of the haulage tunnel produced the first specimens of rhodochrosite in the early 1980s. These were from vugs containing pale, moss-green quartz clusters (colored by inclusions of chlorite) associated with rosettes of pale pink rhodochrosite. The pink-and-green combination was stunning and immediately raised collector awareness of the Grizzly Bear Mine.

At about the same time and in the general vicinity of the chloritic quartz vugs, Grizzly Bear miner Ted Hurr found a pocket of lustrous pyrite crystals covered with bright pink rhodochrosite rhombohedra to 8.5 millimeters. Further along the Zanett Tunnel, miners found large vugs containing barrel-shape white quartz crystals associated with small rhombohedra of rhodochrosite. Plates measured from 33 to 66 centimeters across, although the rhodochrosite crystals on these plates were all less than 2.5 centimeters on edge. Unfortunately, some of the crystals on the largest plate were damaged as the plate was forced out of the too narrow opening of the vug.

Quite a few rhodochrosite rhombohedra on drusy quartz matrix were also removed from a large watercourse in the Zanett Tunnel during a midnight mining effort when the mine was temporarily shut down in 1983. While Grizzly Bear rhodochrosite crystals are generally strawberry-pink, those from the watercourse had been etched and stained dark orange by the naturally acidic, iron-laden water (a result of decomposing pyrite) in the mine. Rumor has it that most of these specimens were placed in a 55-gallon drum containing hydrochloric acid (very damaging to carbonate minerals) in order to remove the heavy iron staining. Several untreated specimens from the watercourse were obtained by the author and were appropriately cleaned.

the deep furrows seen in the hydrochloric-soaked crystals.

As work to drive the 7,000-foot tunnel continued, a "doghouse" was constructed. A doghouse serves multiple purposes, acting as a safety refuge, a lunchroom, and a warming room. High up on the rock face at the edge of the doghouse, Pete Klein (one of the operating partners of ZMK Mine Construction, which was driving the tunnel) found a vug containing a large, quartz-encrusted rhodochrosite rhombohedron. Looking a lot like a wedge of wedding cake, this crystal measures 7 centimeters on edge. It is among the largest of the documented rhodochrosite crystals from the San Juans. Pete Klein later gifted the specimen to the author, and it is a centerpiece of her collection.

As the tunnel was extended under Bear Creek and entered the area of the historic workings, more rhodochrosite-bearing vugs were encountered. In general, these rhombohedra are cherry-red and on a matrix of massive to crystallized quartz mixed with pyrite, sphalerite, and other sulfides. Good specimens of root beer-color sphalerite crystals associated with small, pale pink rhodochrosite rhombohedra were also found in this area.

The Grizzly Bear Mine ranks second to the Sunnyside Mine in terms of the abundance and quality of its rhodochrosite crystal specimens. Ceasing operation after the work-related accidental death of one of ZMK's long-time employees (and Ouray County Commissioner) Joe Mattivi, the mine has been turned over to the Forest Service and is permanently closed, thus ending 135 years of mining hopes, dreams, frustrations, and tragedies.

Mountain Monarch/Michael Breen

About a mile up the Engineer Pass Jeep road from Highway 550 are the portals of the Mountain Monarch Mine and the Michael Breen (Mickey Breene) Mine. Both are part of the same mine complex, which consists of 16 patented and unpatented mining claims and several additional portals. A moderately successful lead/silver mine, by 1900 the complex consisted of a boarding house, mill, mine office, power plant, and a number of other buildings surrounding the portals. The best hand-sorted ore assayed at 658 ounces of silver per ton.

Although Ouray County brothers Don and Ken Nordlander had been sporadically retrieving rhodochrosite specimens from this mine complex in the late 1970s, the real production took place when the property was leased, for a three-month period in the summer of 1982, by the late mineral dealer Harvey Gordon of Reno, Nevada, and his partner Carl Faddis of Belleview, Washington. The Nordlander brothers were hired to do the mining, and several small pockets containing a total of about 20 fine rhodochrosite specimens were recovered. The story goes that Harvey accidentally shot himself in the foot on the second day of mining and thereafter supervised the operation from crutches.

These specimens look a lot like rhodochrosite specimens from the Sunnyside Mine as well as those from the Pasto Bueno Mine in Ancash, Peru (which was producing some very fine specimens during this period) and some concern has been voiced about the legitimacy of the attribution of specimens to the Mountain Monarch Mine. Having personally collected specimens from the locality, the author can vouch for their legitimacy.

Several characteristics distinguish specimens from each of these localities. The Peruvian rhodochrosite occurs on clusters of somewhat elongated, lustrous, semi-transparent quartz crystals, not on druses of white quartz as at the Mountain Monarch. In addition, Peruvian rhodochrosite crystals are not themselves coated with a light druse of quartz, while perhaps 85 percent of the specimens attributed to Mountain Monarch are so coated.

The Sunnyside Mine rhodochrosite specimens are a bit more difficult to distinguish. However, the Mountain

Monarch specimens are almost all sprinkled with small, twinned chalcopyrite crystals. Similar Sunnyside pieces are never found with chalcopyrite, although they are occasionally found in association with tiny pyrite crystals. Additionally, rhodochrosite from the Sunnyside Mine is rarely encrusted with drusy quartz, but as mentioned above, the majority of those from Mountain Monarch are coated.

Although some of the earliest specimens for sale were attributed to the Mickey Breen Mine, some of the specimens appear to have actually come from the adjacent Mountain Monarch Mine. Also, plates of pale pink rhodochrosite on cubic green fluorite were found in an upper level of the Mountain Monarch, as were a few rhodochrosite specimens on black sphalerite crystals.

In the winter of 1983-84, an avalanche destroyed several buildings at the mine, and a subsequent cave-in near the mouth of the portal blocked access to the specimen-producing areas. There has been no mineral collecting since, although specimens occasionally appear on the market.

Daniel Bonanza Mine, Ouray County

The Daniel Bonanza Mine is high up on the rock benches along the west side of Uncompahgre Gorge south of Ouray. The mine is part of the Hayden-Comel claim group. The landmark Sutton Mill was on the site until 1982, when local miscreants accidentally burned it down. The mill was connected by a tramline across the gorge to the Daniel Bonanza and the other mines in the group.

The Daniel Bonanza can be reached by hiking along the benches south of the Mineral Farm Mine and the housing subdivision of the same name. Several avalanche chutes and a lot of downed timber make the long and arduous hike both rugged and hazardous.

There is an old miner's cabin on the benches about ¼ miles south of the Daniel Bonanza that is clearly visible from the highway across the gorge. For years Ouray residents have been placing a clothesline by the cabin along with a sign which reads, "ANTIQUES 9-5." The double-takes of amazed tourists at the Bear Creek scenic overlook are worth the effort put into maintaining this hoax.

The mine consists of two shafts and a connecting tunnel to the deeper shaft (Kelley, 1946). The specimen-bearing vugs are near the bottom of one of the shafts. Although small, the Daniel Bonanza Mine yielded a number of specimens of pink rhodochrosite rhombohedra to about 2 centimeters on edge that are coated with lustrous drusy quartz. Unfortunately, considerable iron staining beneath the quartz diminishes the quality of the color of the rhodochrosite. Cleaning removes most, but not all, of the orange tinge and makes these specimens easily recognizable. The largest of the Daniel Bonanza rhodochrosite crystals is 2.5 centimeters on edge; plates can reach 7.6 centimeters.

Brooklyn Mine, San Juan County

The Brooklyn Mine is several miles north of Silverton, on the mountainside east of the Chattanooga Loop on Highway 550. A dubiously promoted venture, the Brooklyn Mine was the focus of a 1980s Securities and Exchange Commission investigation after it presented overly optimistic geologic assessments.

This mine has several adits and some unstable underground workings. It has produced a number of rhodochrosite specimens, most commonly crystallized crusts on chalcopyrite or pyrite crystals. Rhodochrosite and sphalerite serve as the matrix for a number of gold wires. Some of these specimens were certainly legitimate, but at least one specimen attributed to this mine was examined at the Colorado School of Mines Geology Museum in the late 1980s and was found to have been manufactured: the gold wire was stuck into a crack on a cleavage face of the massive black sphalerite.

The Brooklyn Mine has also produced clusters of pyrite crystals and a number of other locality pieces, but the gold-rhodochrosite combinations are the most desirable. In a news note in *Rocks & Minerals* (1955:1), Ouray mining man

Right:
Rhodochrosite and quartz, 10 cm wide
Idarado Mine, San Miguel County
Colorado School of Mines collection
Jeff Scovil photo

Lower left:
Rhodochrosite and chalcopyrite, 7 cm wide
Brooklyn Mine, San Juan County
Dave Bunk collection, Jesse La Plante photo

Lower right: Rhodochrosite, sphalerite and galena, 3.5 cm tall
Brooklyn Mine, San Juan County
Benjy Kuehling collection
Jesse La Plante photo

Facing page, left:
Rhodochrosite on quartz, 8.75 cm wide
Daniel Bonanza Mine, Ouray County
Barbara L. Muntyan collection
Jesse La Plante photo

Facing page, right: The ore bin at the Mountain Monarch Mine
San Juan County
Daniel Kile photo, 1991

Patrick Fancher stated that he found "strikingly beautiful specimens of pink rhodo, black sphalerite, and pyrite from the Brooklyn Mine." He also mentioned gold wires accompanying some of these pieces.

Other Occurrences

There are many other occurrences of rhodochrosite from the San Juan Mountains. A handful of these localities produced limited numbers of crystallized specimens. The **Alaska Mine** in San Juan County produced a few elongated quartz crystals with crusts of tiny pink rhodochrosite crystals. Robert Stoufer collected a number of these specimens in the 1980s.

In the wilderness area south of Silverton near **Columbine Lake** and Chicago Peak, brothers Larry and Joe Piekenbrock hiked about 15 miles and came upon a mine dump containing plates of quartz with "rhodochrosite" rhombohedra, which were nearly 5 centimeters on edge and were completely altered to manganese oxide. Unfortunately, the mine portal was flooded, thus preventing access to virgin material.

The replacement orebody at the **Idarado Mine** (see page 46) yielded a few specimens of rhodochrosite. These are generally druses associated with quartz crystals and black sphalerite. The **Camp Bird Mine** (see page 30) also produced a few rhodochrosite specimens. The author has a 5-centimeter calcite rhombohedron with a pale pink rhodochrosite crust coating the lower half of the specimen. Tom Rosemeyer, who began as a mine engineer and retired as manager of the Camp Bird, has a plate of rhodochrosite rhombohedra that is somewhat earthy in appearance.

Massive rhodochrosite is a moderately common gangue mineral in various San Juan mines including the **Dunmore**, **Silver Queen**, **Mountain Queen**, and **Virginius** mines in Ouray County, and the **Golden Fleece**, **Scotia**, and **Gold King** mines in San Juan County. Some of these localities even produced crusts of very small crystals, which are of interest to die-hard locality collectors.

Right: A circa 1910 postcard depicting the "Sunnyside Mill" in Eureka, San Juan County
John Taylor collection

Facing page: Fluorite on rhodochrosite, 6 cm tall
Sunnyside Mine
near Gladstone
San Juan County
Colorado
Bill Larson collection
Jeff Scovil photo

On the Sunnyside

by Jack A. Murphy

On most summer mornings, the Durango-Silverton narrow gauge train leaves Durango (elevation 6,500 feet), picking up speed as it heads north toward the San Juan Mountains. The distinctive gold-color train cars make the 45 mile jaunt to Silverton (elevation 9,305 feet) loaded with tourists and history buffs looking to relive the glory days of San Juan mining when the trains transported nearly everyone and everything to and from Silverton.

The sinuous hills along the wide flood plain of the Animas River just north of Durango are glacial moraines, a reminder of several Pleistocene glacial advances. The region has been ice-free for 15,000 years (Blair, 1996). The train tracks parallel the Animas River along the valley floor for several miles and then begin the winding ascent into the steep and scenic Animas Canyon. Here the oldest rocks of the region, Precambrian granites and metamorphic rocks, are exposed.

Beautiful crystallized minerals from the San Juan mines are prized specimens in museums and private collections. They are known from the earliest days of mining and have contributed to the scientific understanding of the rich gold and silver-bearing sulfide veins in Tertiary volcanic rocks. Amateur mineral magazines since the 1970s provide an outstanding pictorial record of Colorado mineral specimens discovered when the mines were operating. These publications reflect the upsurge of mineral collecting after World War II, a time when many rock and mineral clubs were started, and collectors and dealers began promoting mineral shows and exhibits. Those were the best of times for acquiring Colorado minerals, especially those from the San Juan Mountains.

Gold in quartz came from the Camp Bird, Idarado, and Sunnyside mines, sometimes as attractive small crystals and often in rich, fine-grained samples suitable for use in jewelry (Murphy and Modreski, 2002). The colorful non-metallic minerals from veins near the towns of Silverton, Ouray, and Telluride were particularly abundant and included quartz, calcite, fluorite, and rhodochrosite. Specimens found by the miners were sold surreptitiously; such "high-grading" was grounds for a miner's dismissal. But availability and opportunity for acquiring fabulous and expensive minerals prompted a large number of miners to take their chances while providing additional income and preserving many beautiful specimens.

When I was a boy, my family visited Silverton every summer for many years. We began going in the 1950s, when many mines in the area were active. We became acquainted with some of the locals, and I have fond memories of Edna Frecker, the proprietor of the Grand Imperial Hotel, who had a mining museum in the back of the hotel; the cabinets were filled with minerals and many old photos of the mines, mule trains, and snow slides. Charlie Thomas, a hard-rock miner since the 1920s, was a particularly good friend. He led us on many weekend excursions to old mines; in fact the first time I went underground, I was with Charlie, who was caretaker of a mine near Silverton. Charlie showed us high-grade veins and crystal vugs; and he introduced us to one of the local characters Jess Carey, who had been a cook and caretaker at the Highland Mary Mine boarding house before

it was swept away in a snow slide. Jess displayed minerals and artifacts in his store on Silverton's Blair Street (Marshall and Zanoni, 1996), and he was generous to kids, loading our pockets with chunks of galena and long, sharp quartz crystals.

What jump-started my interest in mineral collecting was meeting the kids from Silverton who sold terrific crystallized specimens from beer flats and cigar boxes next to the railroad tracks when the Durango-Silverton train pulled into town at noon. The train whistle echoed across the Animas River Valley about 15 minutes before it pulled in. That whistle was my cue to hustle down to the tracks and inspect the many boxes filled with crystals before the tourists descended from the train. The kids' dads and granddads worked in the nearby mines, mainly the American Tunnel (at the Sunnyside Mine), which was being driven under the mountains northeast of Gladstone to reach the lower gold-rich workings of the Sunnyside Mine. Beautiful little specimens ranged in price from ten to fifty cents. A quarter would buy a nice 5-by-5-centimeter cluster of galena, pyrite, or sphalerite crystals, rarely with small green fluorite crystals or pink rhodochrosite crystals. Fantastic quartz crystals from the Idarado Mine were also available and included long single crystals, some with an amethyst tint, or hand-size clusters of water-clear crystals.

Early on, miners casually threw the crystals into lunch boxes or powder boxes and roughly removed them from underground. But when demand for specimens began to increase in the 1970s and collectors were willing to pay a premium for undamaged specimens, minerals were more carefully wrapped and packed. Free gold in quartz, being less fragile, could be carried in bags or buckets.

Well-formed small bright red to pink rhodochrosite crystals on white crystallized quartz from the Sunnyside Mine were the most highly sought Colorado specimens in the 1960s and 1970s. This eye-catching combination is sometimes labeled "American Tunnel," but the American Tunnel is an access route into the lower working of the Sunnyside Mine; thus, these specimens should be labeled, "Sunnyside Mine, near Gladstone, San Juan County, Colorado."

Silverton is a great place to study mining history and minerals. A short drive to the east, the valley at Howardsville, originally called Baker's Park, was named for George Howard, one of the men who discovered the Sunnyside Mine in 1872. George lived in a log cabin on the banks of the Animas River into his later years; he died at 77 in a Denver hotel. Looking east from Howardsville, the view of the sheer cliffs of Galena Mountain (13,278 feet) is incredible! The visible quartz veins cutting the darker Oligocene lava flows of Galena Mountain must have been an exhilarating sight for early prospectors into the area.

Galena Mountain is just west of the Continental Divide: on the east side are the headwaters of the Rio Grande River and the Lake City, Creede, and Summitville mining districts. All of these areas, including those around Ouray and Telluride, are in the vast San Juan volcanic field, a sequence of Tertiary volcanic rocks dating from 35 million years ago that covers some 10,000 square miles in southwestern Colorado.

The San Juan Mountains have been the subject of numerous geological investigations beginning after the Civil War. John J. Stevenson (1841–1924), the first official government geologist to visit Baker's Park, arrived in 1873. He was the New York state geologist and contributed to the U.S. Geographical Survey West of the One Hundredth Meridian (part of the George M. Wheeler Expedition, funded by the U.S. Army). Stevenson's description of the arduous trip from Del Norte to the mineral-rich mountains at the headwaters of the Rio Grande and the Animas rivers is a little-told chapter in Colorado mining history. With a military escort, the expedition's pack train followed the Ute Indian trail over the Continental Divide into Cunningham Gulch.

Stevenson described the undeveloped ore deposits near Silverton and recognized the extent of the glaciation. He contributed reports to the mining volume of the Wheeler Survey (Stevenson, 1875) but did not contribute significantly to the understanding of the geology of the San

Right: Frederick Endlich was one of the earliest geologists to survey and map the San Juans. This drawing of Howardsville, near Silverton, appears in his 1876 report of the area.

Facing page, upper: Passenger's-eye view from the Durango & Silverton narrow gauge railroad (Sept. 2007). Photo Jack Brauer/Wide Range Photo LLC

Facing page, lower: Local kids selling minerals when the Durango-Silverton narrow gauge train brings its load of tourists into Silverton. Jack A. Murphy photo circa 1970s

FIG. 3.—METALLIFEROUS VEINS EXPOSED TO VIEW NEAR HOWARDSVILLE, COLORADO.

Juan Mountains. Nonetheless, his descriptions of the rich veins helped to attract the attention of investors and mining speculators (Smith, 2009). Upon his return to the east, Stevenson resumed his career as a geologist, gaining prominence as a professor at New York University.

The next geologist into the area was American-German Frederick M. Endlich (1851–1899). He was with the U.S. Geological and Geographical Survey of the Territories, directed by the inimitable Ferdinand V. Hayden (Foster, 1984, 1994). The Hayden Survey explored and mapped the San Juan region from 1873 to 1874, after the Ute Indians had relinquished the land. In 1879 the U.S. Geological Survey was established under the Department of the Interior to eliminate the duplication of fieldwork between the King, Wheeler, and Hayden surveys (Bartlett, 1963).

Frederick Endlich was well-trained, holding master's and doctorate degrees in geology and mining from German universities. He was the mineralogist, chemist, and resident expert on mining at the Smithsonian Institution from 1872 to 1880. Beginning in 1873, at age 22, Endlich signed with Hayden, working in Colorado and Wyoming and publishing his work in the Hayden Survey Annual Reports. Endlich (1876) described the geology and the progress around Silverton. He made several first ascents of the higher peaks with the topographic mapping team (Foster, 1984). Endlich was among the first to recognize the vast extent of the volcanic rocks across southwestern Colorado and their relation to ore deposits. The only illustration in his 1876 chapter on mining is of Galena Mountain, towering above Howardsville. This attractive lithograph (above) emphasizes the metallic appearance of the veins.

When Endlich visited the prospector camp of Eureka late in the summer of 1874, the Sunnyside Mine was not yet developed, and he did not explore the remote area high above timberline that would become the mine. The distinctive, high-grade gold ore in large outcropping veins was not described until 1901 (Ransome, 1901). Neither rhodochrosite nor rhodonite from the San Juans appeared in

Endlich's first list of Colorado minerals in the Hayden Ninth Annual Report (1877); however, "rhodonite" from "Eureka Gulch near Howardsville" was listed the following year in the Tenth Annual Report.

Eureka, the picturesque 1860s gold panning camp on the upper Animas River, was well known as the company town for the Sunnyside mine and mill in the early 1900s. Mine access and various buildings, including a large boarding house, are 3 rugged miles up Eureka Gulch. The complex was built next to the picturesque, small Lake Emma. Many tons of food, coal, and equipment, including personnel, rode in open buckets up and down the steep mountain from mine to mill via a long aerial tramway. Tramways also transported ore from the mine down to the mill.

Eureka is a good spot for studying the geology of the San Juans. The outcrops of the Oligocene-age Picayune megabreccia are exposed in the cliffs of the canyon north of Eureka. This intracaldera rock member of the Sapinero Mesa Tuff erupted from the San Juan and Uncompahgre calderas some 28 million years ago (Yager and Bove, 2002). The megabreccia are rocks that slumped into the collapsing San Juan Caldera and underwent multiple stages of mineralization, ultimately forming the veins in the Sunnyside Mine and adjacent areas (Elllingson, 1996).

Charlie Thomas showed my Dad and me the old Sunnyside mill at Eureka in the early 1960s. Many tourists fondly remember the small frame buildings on the flats next to the Animas River. Rockhounds have long scoured the mine dumps here in search of the hard, massive pink material historically referred to as "rhodonite"; however, this mineral has since been shown to be pyroxmangite, $MnSiO_3$ (Casadevall and Ohmoto, 1977).

For lapidary enthusiasts, Sunnyside pyroxmangite is a true Colorado gemstone, most highly prized when found in association with free gold (Murphy and Modreski, 2002). The glory days of rockhounding at Eureka are described in a short report by Mr. Marion A. Speer, an old timer who worked at the mine when the operator was the American Smelting and Refining Company (Wolle, 1977). Speer described how he used to pick pieces of "rhodonite" out of the ore buckets as they came down from the mine and then threw them over the cliff to the canyon below. Speer related that it was from this "hide-away-place" that, with the help of the Silverton Postmaster and his Jeep, he ultimately recovered all of the fabulous pink Fourth of July vein material that he had.

Casadevall and Ohmoto (1977) did the modern geochemical work on the orebodies of the Sunnyside Mine. At the time, the mine was setting records as Colorado's top gold producer. Under the management of Allan Bird, the mine yielded 700 tons per day of gold-silver-copper-lead-zinc-cadmium ore. Casadevall and Ohmoto advanced the geological interpretation of the ore deposits. Theories about the formation of huge calderas across the San Juan volcanic field were established by Wilbur Burbank, Robert Luedke, Tom Steven, Peter Lipman, Fred Barker, and others at the U.S. Geological Survey (Steven et al., 1974).

Today the rich Sunnyside orebodies are understood to have been incorporated as part of a down-faulted block (the Eureka graben), which formed during the resurgent doming of the San Juan-Uncompahgre calderas. Six main episodes of mineralization followed between 13 and 16.6 million years ago that deposited ores including pyrite-quartz (5 percent), banded quartz-sulfide (30 percent), massive galena-sphalerite-chalcopyrite-bornite-hematite (40 percent), gold-telluride-quartz (1 to 2 percent), manganese (20 percent), and quartz-fluorite-carbonate-sulfate (5 percent).

The model for ore deposition developed by Casadevall and Ohmoto (1977) showed that the principal elements incorporated in the minerals were gathered by north flowing meteoric waters, which passed through the underlying Precambrian, Paleozoic, and Mesozoic rocks on the outer margins of the Silverton Caldera. These solutions were channeled through fissures; the heated fluids leached metals from adjacent wall rocks and redeposited them by precipitation in the fault systems of the Eureka graben.

Pyroxmangite is one of six manganese silicates in the Sunnyside ore; most of it is massive, identified from the veins in the peaks surrounding Lake Emma. These species—alleghanyite, tephroite, friedelite, helvite, alabandite, and "rhodondite" (pyroxmangite)—were originally identified and described in thin sections by U.S. Geological Survey's Wilbur Burbank, (1933) from a few specimens collected on the mine dump at Eureka.

When collecting in the Sunnyside Mine in 1976, I concentrated on locating some of the manganese-rich zones, hoping to find vuggy material where rare helvite and

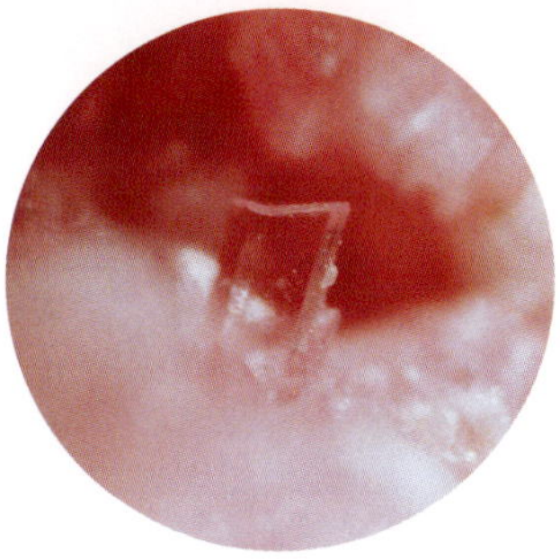

Above: Sunnyside pyroxmangite lapidary work
The slab in the center is 6 cm tall.
Tom Hughes collection, Jesse La Plante photo

Inset left: Quartz on tetrahedrite
(5.3x magnification)
The quartz crystal is 1.9 mm long.
Jack Murphy collection, Dan Behnke photo

Inset right: Pyroxmangite
(13.3x magnification)
The crystal is 0.56 mm long.
Jack A. Murphy collection, Dan Behnke photo

Below: Helvite, 1 cm wide
collected by William Maguire
Dave Bunk collection, Jesse La Plante photo

Facing page: "View of the Tramway at the Sunnyside mines, Eureka, Colo." circa 1910 postcard
Tom Rosemeyer collection

pyroxmangite might have formed euhedral crystals. Helvite, $Mn_4Be_3(SiO_4)_3S$, is one of the most uncommon minerals from the mine, but tiny, prismatic, lemon-yellow crystals were collected and were reported by Rosemeyer (1988). And I found beautiful, rose-red, prismatic microcrystals of pyroxmangite in the 3660 stope on F level, the same locality for the bladed rhodochrosite described below. Colorless, microscopic, scalenohedral calcite crystals are associated with this mineralization but were unreported in earlier publications.

As far as I know, no new manganese silicates have been added to Burbank's suite; however, kutnohorite, $Ca(Mn,Mg,Fe)(CO_3)_2$, has since been recognized. It was identified by Pete Modreski of the U.S. Geological Survey in Denver from material that I collected at the Sunnyside Mine in 1976. This species, the manganese-rich end member of the dolomite group, occurs as minute crystals (some 20-30 microns wide) on blades of the rose-pink pyroxmangite crystals (Eckel et al., 1997).

Through the years, the Sunnyside Mine had its highs and lows, many of the latter resulting in mine closures and loss of jobs. In 1917, Judge John H. Terry, his sons William and Joe, and his daughter Ellen turned the family-operated mine over to the U.S. Mining, Smelting and Refining Company. Illness, death, and low-grade ore during a mining depression

Sunnyside Eye Candy

Upper left: Rhodochrosite, 7 cm tall
Dave Bunk collection, Jesse La Plante photo

Upper right: Rhodochrosite on sphalerite, 6 cm wide
Collector's Edge, Inc. specimen, Jeff Scovil photo

Left: The Postage Stamp Rhodochrosite was purchased in 1966 from Martin Ehrmann for $400.00. The specimen was depicted on one of the four 1974 Mineral Heritage stamps issued on 13 June 1974 at the National Gem and Mineral Show held in Lincoln, Nebraska.

Smithsonian Institution collection (NMNH R16834)
Chip Clark photo

Upper left: Barite on rhodochrosite
2.8 cm wide
Dave Bunk collection
Jeff Scovil photo

Upper right: Rhodochrosite on quartz pseudomorph after anhydrite
3.7 cm wide
Benjy Kuehling collection
Jeff Scovil photo

Left: Gypsum and rhodochrosite
7.7 cm wide
Benjy Kuehling collection
Jeff Scovil photo

Right: Old-time Idaho Springs, Colorado, collector Gordon Nedblake was the first known owner of this 10.4 cm wide specimen of rhodochrosite on quartz from the Sunnyside Mine. Mike Gunnell acquired it from Gordon and sold it to a young Joel Bartsch, who sold his collection in the late 1980s.

Bill Larson collection
Jeff Scovil photo

in the early 1920s forced the mine to shut down. But the Sunnyside had rich ore reserves and a 40-foot-wide gold vein, so when it reopened later in the decade with new development work done on the Washington incline shaft, it was one of the few producing mines in the San Juans.

By 1927 the Sunnyside was the first mine in Colorado to reach a peak production of 1,000 tons of ore per day (Bird, 1986). By 1929, a new 10,000-foot-long tunnel was being planned from Midway, about a mile above the mill at Eureka, to access the underground workings. A tunnel from Midway would have reduced the expenses related to mining ore below F level. It would have likewise eliminated the need for the boarding house and surface buildings at Lake Emma. Had the plans for this tunnel come to fruition, it surely would have been a rich source of mineral specimens; however, the stock market crash in 1929 and the subsequent Great Depression caused the mine to close. And as Allan Bird (1986) noted, "the closing of the Sunnyside was the death knell for the town of Eureka."

In the summer of 1976, I traveled to most of the producing mines in Colorado, hoping to gain access to a mine in order to collect an aesthetic crystal pocket that could be reconstructed for display at the Denver Museum, where I was Curator of Geology. Allan Bird was the general manager of the Sunnyside Mine from 1971–1977. Under his supervision, the mine went from bust to boom. Allan Bird granted my request to collect on the condition that my activities did not interfere with the business of the mine. Later I collected additional crystal vugs from other mining properties: one of quartz crystals from the Idarado Mine, the other a pyrite vug from the Eagle Mine at Gilman, Eagle County. These were reconstructed and are on exhibit at the National Mining Hall of Fame & Museum in Leadville, Colorado (Murphy, 2003).

The search for specimens in the complex and dirty older workings of the Sunnyside was not productive. I had opportunities to examine the freshly excavated rock after the miners had drilled and blasted on a day shift, but this meant riding into the mine through the American Tunnel on the first tram in the morning. This got me into the active stopes early to hopefully find some nice crystals.

With the help of the geology staff at the mine, who showed me around, I got lucky. One morning we scrambled into the 3660 stope on F level in time to see a pile of tan debris on the muck pile that had fallen out of an open hole—a crystal cavity—in the roof! The walls of the pocket were covered with white, tan, and pink crystals! We got a ladder and anxiously crawled into the 7-foot wide hole. It was a well-developed cavity adjacent to a wide sulfide vein. It sloped at an angle of about 45 degrees away from the vein and it was approximately 10 feet long.

The pocket was lined with fine-grained, pink

rhodochrosite and manganocalcite crystals and small, delicate, radial clusters of clear quartz. This was not something I could reconstruct as a crystal cavity exhibit, but it did contain unusual crystal growths; we collected numerous specimens, most significantly 2- to 4-millimeter-long bladed rhodochrosite crystals. In addition, Pete Modreski identified a unique combination: microscopic garnet crystals on pyroxmangite, described and illustrated in Eckel et al. (1997).

I had hoped to return to collect at the Sunnyside Mine, but on Sunday, June 4, 1978, the waters of Lake Emma broke through the upper levels of the mine (see page 81). Miraculously no one was working that day, and there were no human casualties. But the disastrous flood of mud and water put an end to my plans to revisit the mine.

One summer, as Charlie Thomas and I were leaving Gladstone on the Cement Creek Road in Charlie Thomas' pickup, we met a car carrying enthusiastic tourists at the intersection with the highway. They stopped, waved us over, and asked, "Is this the road to Ouray?" Charlie said "Nope." "Well," they asked, "What's up there?" Charlie looked at me, then leaned over and prophetically said to the people in the car, "There's nothing up there but scenery."

Right: Fluorite on rhodochrosite, 11.5 cm tall
Collector's Edge, Inc. specimen, Jeff Scovil photo

Below: Fluorite and rhodochrosite, 10.5 cm wide
Marty Zinn collection, Jeff Scovil photo

Facing page: Circa 1900 photo of Eureka, Colorado, looking south toward Howardsville; Eureka served the Sunnyside Mine until the Great Depression
William Henry Jackson photo
Library of Congress Prints and Photographs collection

The Right Place, The Right Time

by Gary McWilliams

In November 1971, I hitchhiked north out of Durango. I was traveling then by thumb, seeing the country. I was down to my last dime. I'd never heard of Silverton and hadn't the slightest expectation of ever working in a mine. It was only by chance that I was picked up by a car with a driver who offered, "You might find a job in the town ahead, at the mine where I work." Three days earlier, I had been offered a rough-necking job on the oil rigs outside Farmington. Had I not come down with the flu, I would have taken that—become an oil field worker. Chance or fate takes you where it does, inexplicably and by its own whimsy. As it turned out I loved working in the mine and loved Silverton. I stayed there quite a while.

I had not collected minerals since my childhood in North Carolina. I loved minerals then; had a nice little collection of quartz, feldspar, mica, and pyrite. I even wrote a school paper on my future vocation as a mineralogist. The interest went dormant after the family moved to gray and sedimentary western Pennsylvania, as I took up other pursuits more common to adolescence, as I went on to the university to study not geology but the social sciences. But the force of a long-buried interest, when it resurrects, can be amazing. It hit me on my first day underground at the Sunnyside, maybe as I was picking up rocks that had spilled from an ore train, or maybe as I was gazing at the colorful veins of "rhodonite" (pyroxmangite), fluorite, sulfide minerals, and quartz on a tunnel wall. I remember being awestruck from the beginning.

After a while I started collecting some of the nicer samples I found. I learned that virtually all of the miners had little collections at home. The rocks I fancied most were crystals. The mine operators did not care that I took them—they had little or no metal content, were just waste rock. I saved them without cognizance of their monetary value or that a market even existed for them. But soon enough, I discovered rock shops: in Silverton there was Ralph Daily and Wiley Carmack and in nearby Ouray, Benjy Kuehling. Soon enough too, I heard of the Tucson mineral show. In 1973 I went to see what it was about, and sold my first crystals.

I was not the first to collect minerals for sale from the San Juans. One famous collector was Ed McDole (page 90). He had died in Silverton a year before I arrived in town. What a shame that I missed the man; he was not only a legendary vendor of mineral specimens but also a tramp-miner and, by repute, a great storyteller. I loved hearing the colorful tales of the old-time miners who passed through Silverton. Stories of the old West, they were far better than John Wayne or any dish of Hollywood. Ed lived (and died) in a workingman's boarding house in Silverton, the kind of place, not remotely fancy, then common to all of the mining camps in the West. They say he drove a black Lincoln Continental with a trunk load of minerals kept in dynamite boxes and did his buying and selling with nothing but one-hundred dollar bills.

I worked off and on—more off than on, because I was still doing a lot of traveling—at the Sunnyside from 1972 to 1979. The mine was going full bore then, with numerous working places, with a work force of about three hundred. I never worked there long enough at one time to acquire the skills to become a full-fledged miner—the man at the heart of the operation, the man who drills and blasts the rock. My jobs were always lower rung: carrying supplies, greasing machines, operating a small ore train, core drilling.

Before long, I started buying minerals from my coworkers. The miners were friends and let me know—often on the train exiting the mine at the end of shift—if they carried something special in their lunchboxes or in the bibs

From the Collection of Gary McWilliams

Right:
Rhodochrosite pseudomorph after calcite
7.8 cm wide
Idarado Mine, Telluride District
San Miguel County
Dave Bunk collection (ex Gary McWilliams)
Jeff Scovil photo

Facing page, left:
Pyroxmangite, 1.7 cm tall
Sunnyside Mine, near Gladstone, San Juan County
Dave Bunk collection (ex Gary McWilliams)
Jesse La Plante photo

Facing page, right:
Quartz variety amethyst, crystals to 1.9 cm tall
Idarado Mine, Telluride District
San Miguel County
Dave Bunk collection (ex Gary McWilliams)
Jeff Scovil photo

of their diggers. Crystals were not found equally in all parts of the Sunnyside. Often only one or two working areas would have them. Gene Miller and Rick Ernst had the best stope for rhodochrosite. Gene loved the term "blood red." I got a kick out of his oft-used term because the crystals they found were never blood-red but were pink and only pink—the color of Elvis' Cadillac.

The rhodochrosite crystals were often enhanced by the presence of other crystallized minerals. Some of the most aesthetic were perched atop quartz crystals. Occasionally, the pink rhodochrosite would combine with green fluorite. I loved this color combination—like Christmas. Sometimes they combined with crystallized sulfide minerals. Once little bladed rosettes of rhodochrosite were found embedded within gypsum crystals.

Fluorite, usually green and cubic, was often found at the Sunnyside. It could be quite attractive. Crystals of sulfide minerals were found but were seldom large enough or showy enough to be marketable. Wide veins of "rhodonite" (pyroxmangite) coursed through the mine. Virtually all went to the dumps, but a little was saved for lapidary purposes. Very rarely, elongated, translucent crystals of pyroxmangite were found. These were purplish red.

Gold specimens were also collected. When miners high-graded, they generally took any piece of ore with good gold showing. The majority of this was bought by gold buyers who passed though, who crushed it for the bullion. Some of the better was saved and sold to collectors. Such specimens of Sunnyside gold still regularly appear at the various mineral shows. The very best "collector gold" was in the form or leaves or wires. Very rarely, gold crystals (and sometimes tellurides) were found. All were micros.

With a job to do, I could not just roam the mine in search of vugs. I collected crystals when I could. Once, I came across some tiny gold crystals. They were solidly attached within a cavity and I had no tools. As I pondered them, salivating perhaps, I saw a light approaching, someone climbing up the "man-way." As the man got closer, I noted that his "diggers" (the heavy, rubberized-canvas, outer clothing worn underground—the Sunnyside was a wet mine) were brand new, had not a speck of dirt on them. I assumed him to be a new "sampler," one of the young fellows who came and went, who were temporarily hired to collect samples for assay. He had a rock pick/hammer and chisels. "Great," I thought, "Just what I need." I borrowed his tools and, with him standing there, gaping, thinking no doubt that I was just high-grading the gold, I dug out the tiny crystals. As it turned out, this man was not a wet-behind-the-ears sampler but the company's new head engineer. He stayed on to eventually become the mine superintendent. Humorless fellow, he never let me forget the incident.

Gold and rhodochrosite crystals were not found together. They came from entirely different areas of the mine, the rhodochrosite from the lower levels, the gold from the higher. Some years after I left Silverton, when in Alaska, I was approached by a man who claimed to have specimens of gold on rhodochrosite from Silverton, Colorado. "This, I gotta see," said I. He produced for my inspection a box of Sunnyside specimens, as recognizable to me as my sister, of low-grade rhodochrosite on drusy quartz. Gold, in little blebs still showing the melt, had been dribbled over the top. I could not get over the irony of someone in Alaska(!) trying to sell *me* phony specimens from the Sunnyside. I always wondered who manufactured these. The gold had the color of melted down high school class-rings.

Living in Silverton, I came to know the Idarado and Camp Bird miners as well. It was a small town and all socialized together. It was not long before I had started buying their crystals too. Then I started traveling outside of Silverton to the homes of miners living in Ouray, Ridgway, Telluride (I remember it as a mining town), Montrose, Delta, and Durango. Over time, I came to know most of the miners in the San Juans. By lots of asking, by staying on top of who was finding what, I was able to acquire a significant portion of what was being found in the various working mines.

The sheer volume of San Juan crystals that passed through my hands seems, in retrospect, almost unbelievable. I first sold minerals as a dealer at the Tucson Show in 1974. After two years at the Travelodge, I moved to the Desert Inn, Room 201. Two pickup truckloads—beer flats and produce boxes—

went annually to the Tucson show. Throughout the year, I sold truckloads of crystals to Susie Davis of Tucson and later, to Barclay Bitner of Phoenix. At that time, I also made trips around the country selling specimens to rock shops. It combined my two favorite things: rocks and traveling, the former paying for the latter. I always planned my trips to take in the fabled mining camps of the old West: Leadville, Cripple Creek, Wallace, Bishop, Butte, Deadwood, Morenci, Bisbee. In these storied old towns, I sought out the miners—often, I confess, in their watering holes. Many miners had local minerals to sell. Often, they wanted Silverton specimens in trade. Before long I extended my travels to Mexico. Traveling, acquiring beautiful crystals, hearing the stories of the miners, became a lifestyle. I only wish now that I had recorded all that I saw and heard. Once I loaded a U-Haul truck with crystals and rhodonite and drove it from Silverton to Miami, Florida, stopping at all of the rock shops between. I went to Peru for six months with that money. Once, with a pickup truck loaded to the max, I departed Silverton for New York City, where I had a buyer for the lot. My plan was to sell both the minerals and the truck, fly to Europe, and live on a Greek Island (undoubtedly with Melina Mecouri) until all was gone. Alas, fate had it otherwise. I wrecked the truck, and the crystals, in New Orleans. [1]

The previous paragraph bothers me a little because it so emphasizes quantity, as if all of this activity had to do with some bulk commodity like coal or soybeans. I will deal with quality later.

The Idarado produced the most specimens. Because I was a known and regular buyer, many miners, who had before not bothered, were encouraged to bring out crystals. The Idarado then had about four hundred miners and dozens of working places. One of the mine's geologists, Jack Crawford, once gave me a crystal tour. I saw stopes that were full of cavities, their rock walls looking like Swiss cheese. Some of the cavities were huge: I walked into one! The number of vugs, along with the number of employees with access, explains how this single mine produced so many crystals.

I spent a great deal of time cleaning and trimming. In the winters before the Tucson show, I remember cleaning crystals in the backyard of my home, with rubber buckets full of water, acid, and crystals sitting on maybe three-feet of snow. I remember another crystal-cleaning scene. This one took place in Montrose where, at a car wash, I had laid out newly purchased plates of quartz crystals to clean with high-pressure water. Cowboys, always there washing the range mud off their pickups, would mill and ponder at the sight. I recall one insisting to his buddies that they were "petrified ice."

The crystals I sold at Tucson spread around the world. There were many Japanese buyers then as well as Europeans. A few years ago, I had the great privilege of visiting Idar Oberstein in Germany. What a thrill to see the large plates of quartz and calcite crystals on display that had been acquired from me back in the mid-1970s.

Some of the miners took great care to protect the crystals they found. Johnny Davis, who lived in Montrose and worked at the Camp Bird, brought out crystals in impeccable condition. Among much else, he produced some of the first and finest Japanese twins to come from here.

Other miners took less care, as if the direct relationship between money received and condition of crystals never registered. Or perhaps they were only willing to put just so much energy into it. I recall one Idarado miner who brought out a powder box full of crystals a week. They were always just piled on top of each other with no protection for anything. What saved the day, why I bought them anyway, was that most were a calcite-covered quartz. The calcite took the abuse. By dissolving it away and by careful trimming, I could sometimes pull a "silk purse from a sow's ear."

Small mines in the San Juans were always opening and closing. A few produced excellent crystals for their short while. Three come to mind. The Portland (pages 43 and 89) produced beautiful quartz and calcite plates but is best known for lime-green sphalerite. The Grizzly Bear (page 63) produced rhodochrosite crystals that typically came atop green quartz. The Brooklyn Mine (page 66) had beautiful wire and leaf gold in a matrix of black sphalerite. Here, in sharp contrast to the Sunnyside, rhodochrosite and native gold did occur together. Some of the most amazing specimens to ever come out of the San Juans came from this mine. Unfortunately they were few. Perhaps the best was a thumbnail, about a half inch high. In incredibly aesthetic display, it was a thick rams-horn of bright native gold that stood erect from a silvery galena crystal. Bladed, pink rhodochrosite crystals were on three sides and on top. I remember showing this piece to Dr. Frederick Pough, of the American Museum of Natural History, who in 1979, visited Silverton to see my collection. I think that fine man, if he were here today, would not object to my using the word "drool" to describe his reaction.

Of course, I retained the best of all that I acquired for my own personal collection. How I loved this collection. You mineral collectors will understand. But, I was also a young man then and had another love: traveling and seeing the world. Such is my peculiar quirk. It is hard to travel as I liked to then, as unencumbered as the breeze, when you possess—and are possessed by—a rock collection. They're not stamps, after all

In Alaska, I bought a live-aboard boat. I sold the collection in 1979 to finance its rebuild. The buyer, Tom Burrell, a Silverton business man, planned to use the collection as the center piece of a mineral museum he proposed to build in town. I was pleased at the notion of the collection being on public display. Unfortunately, he changed his mind. Far more unfortunately, while Tom Burrell owned the collection, all of the gold specimens, including the beauty from the Brooklyn described above, were stolen.

The collection was thence sold again. Most of it now resides in Denver. Specimens from my old collection have appeared as photographs in various publications on the minerals of the San Juans, this one included (see pages 30, 37, 53, 78 and 79). I was simply lucky to have been there to see all of these Mona Lisas of the natural world, to have been a link in their coming to public appreciation, to have been at the right place, at the right time.

[1] See story "Hot Coffee" in author's collection of travel stories: *Hot Coffee and Other Wild Goose Tales.*

A circa 1921, predisaster view of Lake Emma and some of the Sunnyside mine buildings
Benjy Kuehling collection

Disaster at the Sunnyside

excerpted with permission from *Hot Coffee and Other Wild Goose Tales*
by Gary McWilliams

From 1971 to 1977, I worked at the Sunnyside four times. At no time was I ever a "miner," as the term is used by people in the profession; a "miner," the man at the heart of the process, drills and breaks rock. I never stayed long enough or accumulated enough experience, to be promoted to that esteemed status. The various jobs I had were all supportive to that main activity. One job was as the second man of a two-man core-drilling team. The mine geologists were constantly searching for new orebodies and the core drill was the primary method used. Our machine would be set up in some out-of-the-way place in the mine and we would drill in the precise direction given to us by the mine's engineers. Most of the holes went a hundred feet or so, but some went as far as three hundred. The cutting bit is hollow so that a rock core is collected inside as it moves forward, as new sections are added at the stem. The geologists would eagerly examine this core to see what was ahead in that direction. Occasionally, rich veins were intersected—good news for everybody.

Swede Gunderson was the lead man of our team. On my first day with him, I asked over lunch, "How often do we need to lubricate these drills?"

Swede, opening a little bottle of a prescription medicine, "Take these pills? Yes, I take the damn things for my arthritis." I thought it was a peculiar quirk—something odd about the old man—that his responses to my questions seldom related to what was being asked. Since he wore no hearing aids, since he did not mention his hearing disability, it took a while to realize that Swede heard but little of what was said to him. I learned, if it was important, I'd better shout.[1]

I worked but five months at the mine this time. I quit a little earlier than I might have because my lead man quit. Swede walked out for a specific reason. It had to do with water.

Water is vital to a mine. It is used to cool, lubricate, and flush the drills steels as they bore through rock. At the Sunnyside, water was pumped to the various work stations throughout the mine from sumps on the lower levels. As the

[1] Now, as I write this, I, too, need people to shout. And, I suspect that the major part of my hearing loss had the same origin as Swede's. Without knowing the harm I was doing to myself by not wearing ear protection, without realizing, at the time, that such even existed, I operated a screaming, high-pitch core drill in tightly enclosed mine tunnels. My ears rang every day after work, and have not quit ringing since. No one at the mine wore eye protection, either. Nor were particle masks worn, no matter how thick the air with dynamite smoke, diesel exhaust, and silica-filled rock dust. Many of the older miners had lung problems, and many miners died from what they called the "con," short for consumption, medically known as silicosis. Then, well before OSHA, miners wore no protective gear other than a hardhat and steel-toed boots. Generally, they were not aware of the health risks. To the extent that they were aware of them, they simply accepted them as part of mining. The prevailing attitude was: if you don't like the working conditions, quit. You are responsible for your own health. No one makes you work here.

miners worked ever upward on the veins, getting water to the drills became increasingly difficult. One day, the front office announced that the mine's water problems were soon to be history; that the company engineers had come up with a solution. Much of the mine, they pointed out, lies directly beneath Lake Emma. Why pump water uphill, why fight gravity, when "nature hath provided." We will do the obvious, they said, and tap the lake. A hole drilled upward from the top level of the mine would nicely do. "Lots of water there," and, with the elevation drop, "good pressure too."

Swede and I got the assignment. Swede did not like the idea. He mulled it over and over as we were concluding our prior job. I strained to remember my high school physics. What was Bernoulli's Law? What did I once feed back to Dr. Little about the variables and effects of water volume, distance of fall, and the confinement of space? Swede never studied physics, never went to high school at all, but it seemed to his practical mind that boring into the bottom of a lake from below, while "granted, a clever idea from the standpoint of the college boys sitting in the office," might be less smart for the men drilling from beneath. He visualized the pressure pushing our drill steels back at us with such force "as to whip them into spaghetti and to make us into sauce." Thinking about TV images of water cannon mowing down political protesters, thinking that that amount of water pressure had to be minuscule compared to what could blow through that two-inch hole directly over our heads, I sided with Swede.

After Swede and I walked out, and the mine's other core drilling team declared that they would do the same, the company canned the idea. But they got all the lake's water in another way, in a way most unreckoned by the college boys in the office.

At that time, the mine's richest orebody was the Spur vein. The gold in the higher working places of the Spur was amazing to see. It was the big, chunky stuff that made men a little crazy. The company, maybe made a little crazy too, wanted that gold. As the miners worked upward on the Spur, they encountered more and more ground water. Miners at the Sunnyside were accustomed to constant drips and little streams coming out of fault fractures, but this was excessive. That they were drilling and blasting under Lake Emma, no doubt, played on their mind, made them nervous. They brought their concern to the attention of the shift boss, and he up-the-ladder to the mine foreman, and he up-the-ladder to the mine superintendent. In reply, they were assured that all was safe; that the engineers had checked and rechecked their calculations; that they still had ninety feet of rock between them and the lake bottom. Ninety feet of "back"—in mining terminology—is considered a completely safe margin. The engineers, they were told, had gone out on the lake in a raft and precisely measured its depth with lead lines. They also used a hand-held electronic depth finder. The lake was only twenty-five feet deep at its deepest point. And that deepest point was ninety feet above. "No reason to worry, there is lots of working room."

The miners reluctantly agreed to go on. As they went higher still, the gold got even richer—reported as not just patches, but as a solid streak in the stope walls. But then, they encountered something else: an open space, albeit plugged with mud and loose rocks, in the otherwise solid rock. About twenty feet long and four feet wide, it was, they at first thought, an unusually large cavity. But it had no top, no upper end. How high was it going to extend? To the lake? The miners did not like the looks of it at all.

Standard Metals was owned by New York capitalists who had not the slightest interest in the health or welfare of anything beyond their bottom line—none had ever visited Silverton. It was an attitude that in many ways cost them dearly. But, on the ground, or under it as the case happened to be, the mine was operated by responsible men who were part of the community. All studied the situation. A great deal was at stake. One, writing after the fact, said they "sweat bullets." The local management ended up agreeing with the miners; that there was real danger. Work would be discontinued in that stope until they could drain the lake, which was fully on company property, the following summer. The miners agreed to make one final cut, since the prep work had already been done.

But, the next day, when the time came, the miners said they had changed their minds. They refused to re-enter the area. The wife of one of the men had had a dream the night before—a premonition of the lake's breaking through and flooding the entire mine. She made her husband promise that he would not go back. His two workmates—they had been partners for years—said they had to go along. The shift boss moved the crew to another stope.

That was on a Friday night. Only a skeleton crew worked on Saturday. On Sunday, the only day that the mine was completely shut down, the only day when no one, not a single person, was inside, the roof of that stope collapsed and all of Lake Emma poured through the hole. Indeed with considerable pressure—the engineers had that right—the water, and the bottom mire it carried, flushed through the mine from top to bottom. The pressure of its fall was so tremendous that it flattened a twenty-ton diesel locomotive on the main haulage level. All the machines, all the timbers, all the rail, all the piping—everything man had put into that hole—was crushed or stripped out as the whole of the lake suddenly fell 1,500 feet and blew out through the mine's lower portal. The whole lake emptied in a mere couple of hours. It was one gigantic toilet flush.

Had it not occurred on a Sunday, it would have been one of America's worst mine disasters. When it was over, the sediments from the lake bottom filled every square inch of the mine, plugging it solid. The men underground would have had no place to escape. The average daily workforce of the mine was 130 men. Buried miners would have left only their casts, bones, and clothes, in the hardening—desiccating as the water drained out—clay-like mud.

On the surface, Lake Emma did not cover a big area, maybe a couple of acres. When the engineers measured the depth, they measured down to the bottom of the water, to where they hit something they unwittingly interpreted as the lake's bottom. What they hit was not the bottom, not solid rock, but the top layer of a great accumulation of loose, unconsolidated sediments—the thicker stuff that drained into the hole, that plugged the mine after the first flush of water passed through.

In a book written about the Sunnyside, a former mine

superintendent lays the blame on a very localized quirk of the geology. He explains that the lake was formed by the gouging of a retreating glacier; that the sediments at its bottom were glacially deposited, and that the moving glacier had pried loose a twenty-foot wide rock sliver from the downhill side of the hard-quartz Spur vein.[2] This giant sliver extended sixty feet deep and opened between it and its mother Spur vein an unknown, and unknowable, passageway to the lake. Filled with mud and debris, this fracture zone, as bad luck would have it, was encountered by the miners working upward, and its bottom plug blasted loose.

With a huge insurance settlement and money borrowed on remaining gold reserves, the company spent the next two years cleaning out the solid pack of hardened mud. All the equipment had to be replaced. The prime motivation to reopen the mine was to access that known gold at the top of the Spur vein. The lake, of course, had ceased to be an issue, had ceased to be entirely. The mine operated for thirteen more years. With metal prices low and its best ore mined out, the Sunnyside closed for good in 1991.

The Remains of Lake Emma
The scene above the Sunnyside Mine near Silverton
San Juan County, after Lake Emma drained through the mine.
Paul Carrara/USGS photo, September 12, 1979

EPILOGUE BY ALLAN YOUNG: *The main reason that the amount of pillar remaining below the lake bottom was underestimated was indeed because there was a crevice alongside the vein that escaped detection. This crevice was full of glacial rock flour, as was the bottom of the lake, which made it even more difficult to determine its true profile. This material, as well as contents of stopes, raises and drifts, was flushed through the mine in a very short time and into Cement Creek and the Animas River. The resulting downstream environmental impact was, at least initially, a big story. The rush of water took out all the bridges on Cement Creek leading to the mine, and reportedly turned the Animas River black all the way downstream to Durango. Drinking water supplies as far away as Farmington, New Mexico, were even threatened. Echo Bay Mines purchased the mine in the mid-1980s, and this company began an aggressive development program to increase mine production. I arrived at the mine in May 1987, nine years after the deluge, as the new Mine Manager. During my first trip underground, I visited the site of the Emma Lake breakthrough. The mine had hired a contractor to drill and blast the pillar that remained between the mine and the lake. A streak of fine gold in quartz about four inches wide could easily be seen running most of the length of the pillar. It was a surrealistic scene, with icicles hanging from the blastholes, sunlight streaming through the hole that was above and beyond the protective pillar we were working under, and every so often, a chunk of rock or ice the size of a sofa would come whistling down past us into the partly empty stope below. The mine finally did get its gold, and because of this the company made a good profit that summer and fall. The mine's profitable days were over by the late 1980s.*

[2] Substantiating this theory, large blocks of ancient glacial ice were found at the bottom of the drained lake.

Hübnerite from the Adams Mine

by Daniel Kile

Hübnerite, the manganese end-member of the hübnerite-wolframite-ferberite series of tungstates is among the more noteworthy minerals discovered in the San Juan mountains in the past 100 plus years of exploration. Specimens from the Adams Mine rank it as one of the finest localities for this species in North America; overshadowed in worldwide importance only by hübnerite from Pasto Bueno, Peru.

The Adams Mine, located north of Silverton, is east of Cement Creek and on the western slope of Bonita Peak at an elevation of 11,300 feet. Hübnerite was identified from this mine in the early 1880s, with production recorded by 1899 (Belser, 1956). Early prospecting apparently focused on ores other than tungsten, as evidenced by F.L. Ransome's 1901 report of hübnerite being discarded on the dump, but that nonetheless notes cabinet specimens from the mine "forming striking radial clusters of reddish-brown crystals embedded in quartz and pale-green fluorite." There was a substantial production of tungsten ores during World War I, and minor production in 1955 (Cobb, 1960); the mine has apparently been idle since that time.

The Adams is developed by a haulage tunnel driven about 220 feet to where it perpendicularly intersects the vein. About 160 feet has been drifted along the vein, and the workings include a 90-foot raise, complete with a decaying wooden ladder. A deep surface cut, now mostly caved, is exposed above these workings. The Bureau of Mined Land Reclamation has closed the mine entrance.

The vein, in an altered andesite, is reported to be 3 to 4 feet wide, with hübnerite occurring in bands to 18 inches wide (Belser, 1956) or in irregular nests of radial crystals embedded in quartz and fluorite (Ransome, 1901); fine-grained pyrite and fluorite are associated with the hübnerite in the massive quartz gangue. The author's examination of the underground workings in the early 2000s did not reveal any remaining tungsten ore.

Good specimens from this mine are scarce. The better specimens are composed of lustrous, subparallel, bladed and well-terminated freestanding crystals on matrix. Individual crystals may exceed 5 centimeters in length and reach 19

Above: Hübnerite, 11.6 cm wide
Adams Mine, Bonita Peak, San Juan County
Canadian Museum of Nature collection
(CMNMC57515, ex Bill Pinch), Michael Bainbridge photo

Upper left: Hübnerite, 3 cm wide
Yukon Tunnel, San Juan County
Dave Bunk Minerals specimen, Jesse La Plante photo

Lower left: Lower part of the main dump at the Adams mine, looking toward Cement Creek, Daniel Kile photo

Right: Hübnerite, 7.5 cm tall
Adams Mine, Bonita Peak, San Juan County
Dave Bunk collection, Jeff Scovil photo

millimeters wide; one specimen exhibits a well-terminated, 7.6-centimeter-long crystal. Some crystals retain sharp faces on their terminations, but they often are terminated with irregular and ragged edges. Aggregates to 7.6 centimeters across composed of compact intergrown radiating crystals enclosed in massive quartz have also been noted.

Some of the finest crystals were collected, presumably from underground, by Ed McDole sometime in the 1950s or 1960s (during which time he had taken up residence in a Silverton hotel); it is likely that his finds represent many of the better matrix specimens that are occasionally seen on the market today. More recent exploration of the underground workings in later years has not revealed the exact location of his find, although it has been rumored that it was a large pod collected from the floor of the mine.

An amusing story, related by Gene Schlepp (Bideaux, 2002), tells of McDole's stay at the Desert Inn during an early Tucson Show, after having recently recovered specimens from this mine; needing to etch the enclosing quartz off the specimens with hydrofluoric acid (arguably among the most treacherous chemicals known), he reportedly utilized the hotel room bathtub .... with predictable consequences to both drain and porcelain.

With the underground workings now inaccessible, further prospecting for hübnerite will have to be done on the mine tailings. However, persistent work during the past 20 plus years has yielded some excellent specimens from the dumps. Well-formed crystals exceeding 5 centimeters long on matrix having been collected as have larger aggregates of hübnerite. Nonetheless, the mine dumps have been systematically worked, and productivity has predictably diminished.

As found, many crystals have a drusy quartz coating, while others are completely enclosed in massive quartz. The hübnerite can be etched from quartz with hydrofluoric acid, but caution is due as it is not uncommon for freestanding hübnerite crystals to be partly or even completely replaced by quartz, which would otherwise give quite nice pseudomorphs. An attempt to remove the quartz from such specimens would result in irregularly shaped and unappealing crystals. Specimens appearing on the market usually show hydrofluoric acid etching on the enclosing quartz matrix. There are other mines in the Silverton area that have produced specimen-grade hübnerite, including the **Galtie Boy** (also known as the Galty Boy) in Dry Gulch, the **Yukon Tunnel** east of Cement Creek, the **Ruby Mine** in Maggie Gulch, and in the 1990s, the **Sunnyside Mine** near Gladstone. However, none of these mines have produced the quality or quantity of hübnerite as that which has been recovered from the Adams Mine.

Mineral Collecting in the San Juans: Anecdotes and Advice

by Robert Stoufer

I moved to the San Juans in 1977 and on my arrival began collecting minerals in the prospects around Silverton and later Ouray, but it wasn't until I began collecting with Tom Rosemeyer in about 1982 that I began to find better minerals. Tom had been a geologist and mining engineer at the Camp Bird Mine (see page 30) for many years, and he is an expert at underground exploring and collecting. He, and the many others with whom I have collected, have taught me a great deal. And we have enjoyed some memorable collecting trips.

The San Juans are a great place to live or visit if you enjoy field collecting, mining history or simple mountain scenery. But mineral collecting in the mountains can be hazardous so being prepared and having an understanding of the cultural and natural environment of the San Juans is crucial to a positive experience.

The U.S. Forest Service or the Bureau of Land Management, as well as private mining claim owners, oversee much of the area in the San Juan Mountains. And towns or counties own some as well. Permission is not required to collect minerals on public lands, although the specimens that are found are not supposed to be sold. For privately owned mining claims, permission should obviously be gained before collecting. A little research and a good, up-to-date map are worthwhile investments.

Mountain Hazards

The lowest altitude in the San Juan Triangle is in the area around Ouray (7,800 feet), and some people need several days to acclimate to breathing comfortably in the mountains, as many collecting areas are above 12,000 feet; some are even above 13,000 feet. Altitude sickness, sunburn and other hazards of spending time at high altitudes are real concerns, and collectors should understand the health risks and remedies unique to mountain collecting before venturing out.

Many collecting areas are accessed via 4-wheel drive roads that are very steep and narrow and are often treacherous, especially when they are wet or icy. It can snow or hail in the mountains any time of year. Thunderstorms blow in quickly and lightning and rain can make high-altitude collecting all the more dangerous. Flash floods are possible with any thunderstorm and can potentially close 4-wheel drive roads for a day or longer. It is, therefore, important at all times to carry plenty of potable water as well as emergency food supplies. The 4-wheel drive roads vary in difficulty, and information about the roads in each area can be obtained from the locla Visitor's Centers, as well as from tour companies.

When collecting in areas where mountain goats roam, especially northeast and south of Silverton, it is a good idea to don a hardhat, as the goats like to hang around people and have been known to dislodge rocks and send them bouncing

Right: Quartz, 14.6 cm wide
Gold Lake Mine
Little Giant Basin
San Juan County
Benjy Kuehling collection
Jeff Scovil photo

Below: Quartz, 2.8 cm wide
Meteor pocket
Ouray County
Benjy Kuehling collection
Jeff Scovil photo

Facing page: Tom Rosemeyer (left), Benjy Kuehling (center) and Robert Stoufer collecting on the Mill level of the Ransom Mine, San Juan County, June 14, 1987
Gene Tribbey photo

downhill.

Rocky Mountain marmots abound in the whole San Juan Mountain area and commonly climb up into vehicle engines to gnaw on rubber hoses. More than one collector has had to "coast" downhill after having a radiator hose gnawed through. There are a couple of precautions that may help prevent such a mishap: salt lick, used for cattle, is appealing to marmots and distributing chunks a short distance from your vehicle can divert their attention from vehicle hoses; alternatively, scattering moth balls under your vehicle is thought to repel marmots and other wild animals, which dislike the smell. A recent innovation is to bring a heavy tarp into the mountains, park on top of it, and tie it up around the sides of the vehicle. Temporary wire fencing can also be used to surround a vehicle and keep marmots away.

I recommend that all hikers, including mineral collectors, purchase a Colorado Outdoor Recreation Search & Rescue card. It is valid for 5 years and costs about $12. A card can be purchased at most outdoor sports stores. In case of an accident, the cards insure individuals against the expenses incurred for mountain rescue teams, helicopter extrication, and Sheriff's expenses.

That said, one of the nice things about living in Ouray is its proximity to collecting sites. The Ouray fault, a prominent structure running from literally the city limits of Ouray into the Amphitheater to the east, is host to some of the most important quartz localities in North America. Innumerable pockets have been found along this fault, including the well known **Ohio Mine**, **Crystal Cavern**, **KD area**, **Cliffhanger**, **Laughing Tom**, **OS**, **Watercourse**, **Tree Root**, and **Meteor** and **BK** pockets.

Meteor Pocket

At 4 AM on May 15, 1997, local mineral collector Carl Auclair was driving into Ouray from the north when he saw a meteor streak across the sky that appeared to land near the 4-wheel drive road to the **Portland Mine** (see map page 43 and description below). Arriving at what he thought must have been the area of impact, he began searching for unusual rocks, mainly smoking ones. While he didn't find a meteorite, he did find an unusual knob of quartz, with some small pockets around it. In excavating the chunks of the quartz, he uncovered several small pockets, some with milky quartz crystals to 20 centimeters.

After collecting several similar pockets over a couple of years, Carl brought his friend Benjy Kuehling to the site. While they didn't find much on that trip, Benjy was enticed to come back. He was alone on his next visit to the site on June 17, 1998 and was turning over rocks and prying slabs of the limestone out of the way when he uncovered a clay-filled pocket. He dug through the clay with his "scratcher," a screwdriver with the tip bent at 90 degrees and almost immediately began uncovering large milky quartz crystals, some 10 centimeters in diameter and over 25 centimeters

Above: Sphalerite, calcite and pyrite, 3.5 cm tall
Portland Mine, Ouray County
Brian Kosnar collection (ex Dave Bergman), Jeff Scovil photo

Right: Quartz, 14 cm tall
Ohio Mine, Ouray County
Jim and Imelda Klein collection, Joe Budd photo

Facing page: Chalcopyrite on quartz, center crystal is 2.4 cm wide
Bandora Mine, San Juan County
Collector's Edge, Inc. specimen, Jeff Scovil photo

long. Benjy called Carl to come over and help recover the pocket contents. They spent five days carefully excavating the clay, which had crystals scattered throughout. The emptied "Meteor" pocket was more than 4 feet in diameter and 8 feet deep! The largest crystal is 18 centimeters wide at the base, tapers to a point, and is 69 centimeters long!

Ohio Mine

Many collectors rank the milky quartz from the Ohio Mine as being among the best in the world. Quartz from this locality was first available in the late 1890s. Among the oldest specimens known from this mine are preserved in the Field Museum in Chicago. Specimens have been pictured in publications including *Getting Acquainted with Minerals* (English, 1934) and in *Colorado Gem Trails and Mineral Guide* (Pearl, 1972), but the locality was largely forgotten until Sandy Browne, an employee of Columbine Mineral Shop in Ouray, rediscovered it in the summer of 1983 in the Amphitheater east of town.

Sandy found unusual lustrous milky quartz crystal groups on a small dump; the largest crystals were about 2.5 centimeters in length. He told both me and collector Gene Tribbey about the find, and I went with Tom Rosemeyer to have a look. We discovered the portal still open and a short tunnel that ended in a large solution cavity in the limestone country rock. At first, we thought that someone had broken large crystals off of the roof of the pocket, as there were scars where crystals more than 5 centimeters in diameter had been attached. Milk-chocolate-colored clay covered the floor of the pocket. There was a small pit in the clay near the middle of the pocket, but no crystals were showing; so we figured on that first visit that all of the crystals had been removed from the cavities. But after thinking about all of that clay, we decided to go back and have another look.

A test pit exposed single crystals and a couple of groups of larger crystals. The crystals were unusual in that tiny pyramidal faces covered the prism faces that curved up to the terminations. Multiple generations of milky quartz had overgrown clearer quartz crystals and given them an unusual snow-white appearance. Tiny milky quartz crystals grew preferentially on and perpendicular to three adjacent faces of the 6-sided prisms of the larger crystals. Many broken crystals are "healed," a testament to the fact that the clearer quartz continued to grow after at least one of the events that broke them off of the ceiling of the pocket.

As we continued to collect, we exposed more single crystals and groups, finding large limestone slabs that were partly to completely covered with drusy milky quartz crystals and clay. We also found quartz crystals and groups in the clay-filled areas between the slabs. Tom and I were richly rewarded for our efforts, as were those of many of the collectors who later visited the site.

Portland Mine

Off of the Ouray Fault but still close to town and at the end of the road in the Amphitheater, the Portland Mine was originally mined in the late 1800s to recover base metals and silver, but was reopened in the early 1970s as a specimen mine. During this latter period, hundreds of groups of stubby, translucent and milky, quartz crystals were collected. Around 1980, the mine owners recovered some beautiful groups of clear, green sphalerite crystals to 2.5 centimeters with lustrous small pyritohedral crystals of pyrite, often on milky quartz. They sold the specimens to wholesale mineral dealer Gary McWilliams (page 78).

The mine closed within a year of the sphalerite find, but people continued to venture underground to collect. Near the area that produced the sphalerite, collectors located a pocket on the east wall, or rib, down at the floor level. While raking through the freezing water and broken rock in the pocket, they found hundreds of small (less than 4 centimeters long) doubly terminated, lustrous, blocky milky quartz crystals and groups. Much less common were elongated crystals of light amethyst, exhibiting hexagonal cross sections at the larger ends that gradually taper to triangular cross sections toward the smaller ends. The amethyst crystals are longer than the milky quartz crystals, reaching lengths of 12 centimeters.

In another area of the mine, collectors found a small pocket with distinct scepters of milky quartz (some doubly terminated) up to 8 centimeters long, with preferential overgrowths of drusy yellowish calcite. An area just beyond the sphalerite zone in the west tunnel produced groups of tiny, clear quartz crystals with microscopic, dark green clinochlore crystals and sometimes tiny white dolomite rhombs.

A few years ago, the mine owners put metal vent line pipes underground near the portal and bulldozed the portal shut. They also closed the side entrance, which acted as an escape route and ventilation source. The potential still exists for the recovery of specimens from the Portland Mine, but it would probably take serious mining, including drilling and blasting, to uncover another pocket of good specimens.

Bandora Mine

Back in the summer of 1987, mineral dealers Charles Mochenbacher and Martin Jensen happened to meet local miner Zeke Zanoni in a bar in Silverton. Zeke told them of working at the Bandora Mine up South Mineral Creek west of Silverton and of the quartz and chalcopyrite that he had seen there. Charles and Martin checked out the mine and found Zeke's tale to be true. The pair borrowed a 24-foot extension ladder from Benjy Kuehling, with the understanding that he would have the opportunity to buy any specimens that they recovered.

Charles and Martin collected underground at the Bandora for weeks, filling beer flats with hundreds of specimens, which they then cleaned at Benjy's warehouse in Ouray. Benjy bought the entire lot, except for specimens the two collectors kept for themselves.

Charles and Martin later explored the other open portals of the Bandora. In the "Waterfall" portal along the creek, they found groups of quartz crystals exceeding 30 centimeters in length with clear prismatic crystals 12 centimeters long.

Another portal, south of the tunnel that had produced the chalcopyrite, yielded micro-specimens of cerussite, wulfenite, and pyromorphite, sometimes in combination. Many of the cerussite specimens showed cyclic twinning, with very nice sixling twins.

Later Benjy Kuehling, Gene Tribbey, Tom Rosemeyer, and I went out to the mine to see if Charles and Martin had left anything to collect. We drove to the end of the South Mineral Creek Road and climbed the hillside to the west to the portal of the Bandora. The tunnel was not very long, and toward the working face of the mine, the vein had been stoped upward more than 20 feet. Setting up the same 24-foot extension ladder that Charles and Martin had borrowed on the broken rock on the tunnel floor, we each climbed up to see beautiful chalcopyrite, sphalerite, and quartz groups still in place. We took turns standing on the next to the last rung of the ladder, reaching as high as we could to collect the crystal groups and handing specimens down to the others in the group to wrap and pack. Once back in Ouray, we rinsed off our finds, flipped a coin to see what order we would start our division of the specimens, and then took turns picking from the lot. We each wound up with several flats of specimens.

The specimens consist of simple tetrahedral and interpenetrant twinned, lustrous, golden chalcopyrite crystals to 4 centimeters on edge, although most were in the 2 to 2.5 centimeter range. Lustrous black octahedral sphalerite crystals to 2.5 cm, small cubic galena crystals with a pale gray alteration to anglesite or cerussite, and sometimes a micro-coating of blue linarite crystals (on uncleaned specimens) are associated with the chalcopyrite and prismatic clear quartz crystals. Sometimes the quartz shows sceptering or reverse-sceptering and is frequently partially resorbed.

Subsequent collectors visiting from New Mexico, Colorado, and California also found great specimens at the Bandora, until the Sunnyside Gold Corporation, which owned the mine, blasted the portal shut to prevent further collecting.

Ed McDole:
Mineral Entrepreneur of the San Juans

an appreciation by Robert W. Eveleth

Ed McDole has been gone now for over 40 years, and the author considers himself fortunate to have crossed paths with the legendary mineral collector/dealer during his mining days with Dixilyn Mining Co. in Silverton from 1969-1971. Ed, at this stage of his life, routinely positioned himself just outside of the large doorway between the lobby and the saloon (never in the saloon: despite the legend of the Lemon Hart rum, Ed was diabetic and couldn't touch the stuff) in the Grand Imperial Hotel, (the "GI") from late afternoon until early evening to meet the Silverton miners when they came in to oil their dusty throats. Ed's persona was unmistakable—his trademark long-sleeved white shirt with rolled cuffs, black pants, a permanent cigar firmly lodged to one side of his mouth (in contrast with his pronounced bushy eyebrows) and set off by a Rolex complete with gold nuggets laid into the watchband—that was Silverton's Ed McDole. The author spent many an enjoyable hour talking mines and minerals with the man. From both his location in the GI lobby and his almost daily summertime visits to local mines, untold numbers of classic San Juan mineral specimens found their way into the hands of collectors worldwide.

Ed first showed up in the San Juans about 1963 after meeting local mineral dealers like Wiley Carmack at the Tucson show. Carmack along with Darnell "Zeke" Zanoni were driving a raise at the Sunnyside Mine when they collected some superb rhodochrosite. At the show, McDole took note of Wiley's specimens and then came through Silverton for the first time.

The following year (winter of 1964/65) my colleague John Ross along with Zanoni and Carmack were contract mining at the famous Adams ("Diggings") claims and had collected several flats of those superb huebnerites for which the mine was noted (see page 84). Ed got wind of them and arranged to meet Carmack and Ross in Durango to purchase the lot. These minerals must have convinced Ed that the San Juans held great potential for high-quality specimens as he made Silverton his permanent (summertime) base soon after.

Ed initially engaged in contract mining himself and, by virtue of his mining expertise and knowledge of innovative mining techniques brought from Butte, soon earned the respect of his Silverton colleagues. He showed the locals how to pick up a caved area by "booming" as opposed to the local technique of "spiling" (aka "fore-poling"). But in time, he became more and more focused on the mineral specimen end of the business. Ever the astute businessman, Ed worked out financial agreements with some of the small mine operators/leasers around Silverton that enabled him to be the first to the muck pile after a blast. Very few mineral treasures escaped his attention.

I encountered Ed's long arm of acquisition one day at my employer's Osceola/Pride of the West Mine up Cunningham Gulch. As I approached the portal, I encountered miner Nick Chavez carefully carrying two of the most beautiful chalcopyrite specimens I have yet seen. Each piece was large enough to cover Nick's gigantic palms and the individual chalcopyrites, up to 2 centimeters, were perched on snow-white calcite or quartz crystals. The specimens were remarkable because the surfaces of the chalcopyrite were oxidized to intense hues of red, blue, and green. Even uncleaned they sparkled like gems. I tried to buy one on the spot, but Nick responded "No way. These are going to my buyer downtown." McDole had a long-established reputation of being very generous when buying specimens, often offering more than the miner's asking price because he knew the true value of the material, and he wanted the miners to come back to him with future finds, a fact further illuminated by the following.

In an effort to keep their Howardsville mill fed with ore, Dixilyn Co. leased many properties, in addition to those mentioned above: the Gary Owen, Highland Mary, Ransom, Silver Wing, Henrietta, and others. "Whitey" David held the Dixilyn contract at the Silver Wing, between Eureka and Animas Forks, and while doing development work opened up a vug of tetrahedrite crystals—according to Tom Rosemeyer, the finest occurrence of the species from in the San Juans. The crystals were shiny black metallic, up to 2.5 centimeters on edge, occurring in groups and singles. But before Tom could make an offer, Whitey took them to McDole who purchased the entire lot and sold them to a "closet" collector in Texas. To this day the specimens have been untraced.

Another mine, known for fine enargites, was the Longfellow near Red Mountain Pass. And Ed, having made his "standard" agreement with the leaser, stopped by daily. On this fateful day my colleague John Ross and his partner were attempting to cut samples from the back of a stope when loose muck broke through the timber a few sets below them and sealed off their only means of escape. The duo took care to pull the air hose up with them as they advanced or they would have been dead men in a matter of minutes. Eventually they could hear the slusher start up and they knew help was on the way—the aroma of Ed's cigar announced his presence long before they could see him. Finally when a hole opened up big enough to crawl through, Ross said "the most beautiful sight he ever saw was the moment that big Scots-Irishman poked those bushy eyebrows up through the hole and hollered "are you boys all right up there?" When McDole got an affirmative response he further commented, "[expletive] the air up here is really RANK." Whereupon Ross and his partner shouted, almost in unison "well it was just fine until you poked that stinking cigar up through that hole!"

Ed McDole departed this earth just shy of 58 years on 21 May 1970. He left a legacy seemingly the product of a much longer life. Elsewhere he may have been known as the "Montana Mineral King," but to those of us steeped in Silverton's mining history, he was the "Mineral Entrepreneur of the San Juans."

ACKNOWLEDGMENTS: *The author is grateful to his recently deceased colleague John Ross for introducing him to McDole in 1969. Both the latter now reside not far from each other in the Silverton cemetery. Thanks also to colleagues Tom Rosemeyer, Benjy Kuehling, Frank Montonati, and others who have generously shared their recollections.*

Ransom Mine

Well known mineral collector Gene Tribbey has spent many summers in Ouray, collecting. In the summer of 1985, he was told of a pocket in the Ransom Mine, just above and west of the Eureka town site north of Silverton, and he drafted me to go with him to check out the locality. We donned our underground gear, walked about 3,000 feet into the haulage tunnel, and came to a vertical cleft in the rock on the right side of the tunnel that was partly filled with fallen rock, some of which was covered in clay.

Looking up, we saw cubo-octahedral fluorite crystals, some exceeding 5 centimeters, covering both sides of a pocket some 30 feet above. The sight motivated us to pull the broken pieces out of the bottom of the pocket. They were covered in plastic, light brown clay, and we washed them off in the stream that was running through the tunnel. Some of the pieces were just broken fragments of country rock, some were groups of etched, milky quartz crystals, and some were beautiful groups of fluorite crystals, usually on milky quartz crystals. All of the fluorite crystals had a very thin coating of drusy quartz, which we later removed with hydrofluoric acid. Many of the pale green fluorite crystals had inclusions of micro chalcopyrite crystals.

Gene and I later returned with Tom Rosemeyer, Benjy Kuehling, and an extension ladder to try to collect in the upper part of the pocket. But collecting for 10 to 15 minutes from the top of the ladder exhausted the oxygen at the top of the pocket, forcing the "ladder-man" to come down so that we could fan "fresh" air up into the top of the vug. Those of us on the ground continued to collect the clay-covered pieces at the bottom of the pocket. Some of those specimens from the bottom of the pocket were the best that we collected, showing more relief than showed on the plates that we had collected from the walls of the pocket. The largest plates were 50-by-50 centimeters, with one surface completely coated with 2- to 5-centimeter-wide fluorite crystals.

The year after we had collected in the mine, Sunnyside Gold Corporation blasted the Ransom Mine portal shut, and today it is difficult to tell where the portal was located.

Senorita Mine

In 1983, Tom Rosemeyer and I went underground at the Senorita Mine. The Senorita is one of several mines, including the **Black Girl** and **Newsboy/Slide,** in the Pacquin mining district north of Ouray. In this district feeder veins intersected the Pony Express Limestone and replaced part of the limestone with ore minerals, quartz, and barite. As the mines were driven in sedimentary rocks, rock falls and cave-ins are common; all of the mines in the Pacquin district are privately owned.

Sliding down the partly caved entrance to the mine, we carefully made our way around an interior shaft and were soon seeing vugs in the walls of the mine. One hundred feet or so into the mine, a side tunnel had been driven to the south, and in that area we found pockets up to three or four feet long and a foot high that contained loose specimens of bladed, white barite to 4.5 centimeters on tiny clear drusy quartz crystals covering a dark gray silicified matrix. We collected several flats of specimens, including some unusual calcite crystals, in three visits to the mine.

Above: Fluorite on quartz, left crystal 1 cm tall
Ransom Mine, San Juan County
Benjy Kuehling collection, Jeff Scovil photo

Lower left: Cerussite, center crystal 0.8 mm
Senorita Mine, Ouray County
Tom Rosemeyer collection, Dan Behnke photo (6x)

Lower right: Azurite, crystal 1.7 mm wide
Slide Mine, Ouray County
Tom Rosemeyer collection, Dan Behnke photo (7x)

Further into the mine we found a zone with colorful micro-minerals, most commonly crusts of greenish to white smithsonite and clear hemimorphite crystals. Micro-size examples of malachite, azurite, cerussite, linarite, and other secondary minerals were also common in this zone.

In Closing

There are many other experiences to share after more than 30 years collecting in the mines and prospects of the San Juans. But space does not permit recounts of trips to the **Zuni Mine**, type locality for zunyite, or about our finds at the **Genoa** claim, our many trips into **Little Giant Basin**, to the **Cunningham Gulch** area, or the many other localities in the San Juan Triangle.

The only way to truly experience our mountains, towns and localities is to pay us a visit. When you come, I encourage you to spend a few hours visiting the Historical Society museums in Ouray and Silverton, both of which have San Juan minerals and mining memorabilia on display. If you didn't have good luck in the field, Columbine Mineral Shop on Ouray's Main Street sells minerals from the San Juans as do a number of dealers in Denver and on the web.

And I welcome you to stop in to see us at Buckskin Booksellers in the Beaumont Hotel in Ouray. Our contact information can be found in our ad in the back of this volume. See you in the San Juans!

Ephemera and Memorabilia

Rich with a human history that includes Indians, mining, railroads, and tourists, the San Juans have generated an array of collectibles that extends beyond its prized mineral specimens. Vintage and sometimes rare prints and postcards pepper this volume, and buffs of all sorts vie for the rarest of these and other collectibles such as stock certificates; railroad, Indian, and mining artifacts; and Americana. Collectibles related to the San Juans can be found in antique shops, on (or under) the ground, and in private and museum collections, including the historical society museums in Silverton, Ouray, and Telluride.

Above: Instrumental in the early development of the mines and communities in the San Juans, the Russian-born "Pathfinder" Otto Mears (1840–1931) had built hundreds of miles of toll roads through southeastern Colorado before turning his attention to building his first railroad, a 25-mile-long, narrow gauge line from Silverton to Red Mountain Town. Active in business and politics in the Colorado Territory, Mears issued small numbers of rail passes to friends and investors. These were first made of buckskin, but were later cast from silver that is said to have originated in the San Juan mines. This rare and highly desirable pass, hand-engraved with "No. 317," is in the collection of Benjy Kuehling.
Gloria A. Staebler photo

Left: For centuries, silver spoons have been given to commemorate life's milestones, and wealthy travelers have tradionally purchased silver souvenir spoons, making them a natural for the gift shops in the scenic, silver-producing San Juans. These three circa 1900s mining-theme, sterling silver spoons are from Ouray (left), Silverton (center), and Telluride and are in the collection of Tom Hughes.

The spoons sit atop a copy of Thomas "T.A." Rickard's collectible book, *Across the San Juan Mountains*. Rickard was a renowned mining engineer, prolific author, and Colorado's State Geologist from 1895 through 1901. He was inducted into the National Mining Hall of Fame in 1990. *Across the San Juan Mountains* recounts a 400-mile journey by Rickard and three mining engineer friends through the San Juans to examine the mines of the region. It was published in 1903 and was reprinted as part of *Journeys of Observation* in 1907. This copy is in the collection of Dave Bunk.
Jesse La Plante photo

Right: The Western Federation of Miners, formed in Butte, Montana, in 1893, was a militant union that formed locals in many hard rock mining camps in the American west. Uprisings or strikes were routine in the early days of this union. Four such strikes resulted in armed conflicts between Western Federation miners and mine owners in the Colorado towns of Leadville (1895), Cripple Creek (1894 and 1903), and Telluride (1901). The Telluride strike against the Smuggler Union Mine resulted in the shooting deaths of at least two non-union mine workers and injuries to several others at the hands of striking miners. These elaborate ribbons were worn in parades, funeral processions, and presumably union meetings. They are perhaps the most collectible of all union ribbons.

Miners' candlesticks were usually simple and functional, but improvements to the basic design were regularly patented. Ramstead and Johnson were residents of Ouray in May 25, 1909, when they patented their design, which features a folding hook.

Dave Bunk collection
Jesse La Plante photo

Insets: Exonumia was often introduced as a means of trade in remote areas where currency or coins were in short supply. These tokens were issued by businesses and were commonly given to customers in lieu of small change. They were only redeemable with a specific merchant.
Benjy Kuehling collection
Gloria A. Staebler photos

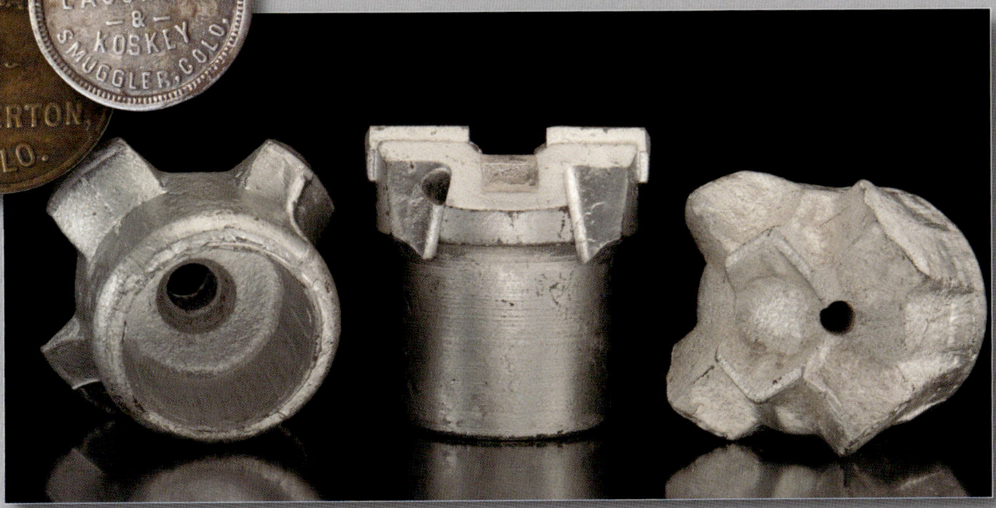

Left: Collectors are commonly on the lookout for mining artifacts while they are prospecting and digging. John Trujillo gathered these drill bits from mine properties in the Ouray area. Other collectors have found miner's lamps, candlesticks, dynamite boxes and all sorts of antique mining implements while underground or turning over dumps. In fact, artifacts are the primary targets of many collectors.
Jesse La Plante photo

References

Bartlett, Richard A. 1963. *Great Surveys of the American West*. Norman: University of Oklahoma Press.

Belser, C. 1956. "Tungsten Potential in the San Juan Area, Ouray, San Juan, and San Miguel Counties, Colorado." *Bureau of Mines Information Circular 7731*.

Bideaux, R.A. 2002. "Ed McDole: Montana Mineral King." *The Mineralogical Record*, 33:71-74.

Bird, Allan, G. 1986. *Silverton Gold, The Story of Colorado's Largest Gold Mine*. Privately published.

Blair, Rob. 1996. "Origin of Landscapes." In *The Western San Juan Mountains, Their Geology, Ecology, and Human History*, edited by Rob Blair, Tom Ann Casey, William H. Romme and Richard N. Ellis, 3-17. Niwot: University Press of Colorado.

Bove, D.J., et al. 2007. Major Styles of Mineralization and Hydrothermal Alteration and Related Solid- and Aqueous-Phase Geochemical Signatures: *U.S. Geological Survey Professional Paper 1651*.

Bove, D.J., et al. 2001. Geochronology and Geology of the Late Oligocene through Miocene Volcanism and Mineralization in the Western San Juan Mountains, Colorado. *U.S. Geological Survey Professional Paper 1642*.

Brown, Delmer L. 1977. *Geology of the Golden Wonder Mine, Hinsdale County, Colorado*. Unpublished report.

Burbank, Wilbur S. 1933. "The Manganese Minerals of the Sunnyside Veins, Eureka Gulch, Colorado." *The American Mineralogist,* 18:513-527.

Burbank, Wilbur S. 1941. "Structural Control of Ore Deposition in the Red Mountain, Sneffels, and Telluride Districts of the San Juan Mountains, Colorado." *Colorado Scientific Society Proceedings,* 14:141-261.

Burbank, Wilbur S., and Robert G. Luedke. 1969. Geology and Ore Deposits of the Eureka and Adjoining Districts, San Juan Mountains, Colorado. *U.S. Geological Survey Professional Paper 535*.

Casadevall, T., and H. Ohmoto. 1977. "Sunnyside Mine, Eureka Mining District, San Juan County, Colorado: Geochemistry of Gold and Base Metal Ore Deposition in a Volcanic Environment." *Economic Geology,* 72:1285-1330.

Cobb, H.S. 1960. *Tungsten Mines of Colorado*. State of Colorado Metal Mining Fund Board.

Daniels, Helen S. 1996. *The Ute Indians of Southwest Colorado*. Lake City: Western Reflections.

Eckel, E.B. 1997. *Minerals of Colorado*, updated and revised by R.R. Cobban, D.S. Collins, E.E. Foord, D.E. Kile, P.J. Modreski, and J.A. Murphy. Golden: Fulcrum Publishing.

Elllingson, Jack A. 1996. "Volcanic Rocks." In *The Western San Juan Mountains, Their Geology, Ecology, and Human History*, edited by Rob Blair, Tom Ann Casey, William H. Romme and Richard N. Ellis. Niwot: University Press of Colorado.

Endlich, F.M. 1876. "Report of F.M. Endlich, S.N.D., 1874, Mines." In *Eighth Annual Report of the United States Geological and Geographical Survey of the Territories Embracing Colorado and Parts of Adjacent Territories Being a Report of Progress of the Exploration for the Year 1874, by F.V. Hayden, U.S. Geologist.* Washington, D.C.: Government Printing Office.

Endlich, F.M. 1877. "Catalogue of the Minerals of Colorado." In *Ninth Annual Report of the United States Geological and Geographical Survey of the Territories Embracing Colorado and Parts of Adjacent Territories Being a Report of the Progress of the Exploration for the Year 1875, by F.V. Hayden, U.S. Geologist*, Chapter IV, Appendix B. Washington, D.C.: Government Printing Office.

Endlich, F.M. 1878. "Mineralogical Report—Catalogue of Minerals Found in Colorado." In *Tenth Annual Report of the United States Geological and Geographical Survey of the Territories Embracing Colorado and Parts of Adjacent Territories Being a Report of the Progress of the Exploration for the Year 1876, by F.V. Hayden, U.S. Geologist.* Washington, D.C.: Government Printing Office.

English, George L. 1934. *Getting Acquainted with Minerals.* New York: McGraw-Hill.

Fetchenhier, S. 1996. "Ore Deposits and Minerals." In *The Western San Juan Mountains, Their Geology, Ecology, and Human History*. Niwot: University Press of Colorado.

Fossett, F. 1879. "Colorado, Its Gold and Silver Mines, Farms and Stock Ranges, and Health and Pleasure Resorts." *Tourist Guide to the Rocky Mountains*. New York: C.G. Crawford Publishing.

Foster, Mike, ed. 1984. *Summits to Reach, An Annotated Edition of Franklin Rhoda's Report on the Topography of the San Juan Country*. Boulder: Pruett Press.

Foster, Mike. 1994. *Strange Genius, The Life of Ferdinand Vandeveer Hayden*. Niwot: Roberts Rinehart Publishers.

Greager, Howard E. 1996. *We Shall Fall as the Leaves*. Self-published.

Ed Eckel (1906–1989) was an icon of Colorado mineralogy with a passion for those found within its borders. Accordingly, over a span of 30 years, in his spare time, he compiled a description of all of the known minerals and their localities, more then 445 species as of 1958, which was published in 1961 as *U.S. Geological Survey Bulletin 1114*. This book became required reading for any who might venture into the field in search of minerals (my original paperback copy is worn almost beyond recognition); its popularity was attested to by the fact the first printing was sold out by 1969.

Since 1958, scores of minerals new to Colorado had been reported, and by the 1970s it had become apparent that an update to the original work was in order. The task was taken up by volunteers within the mineral community, the U.S. Geological Survey, and the Denver Museum of Natural History. A massive amount of literature, both published and unpublished, as well as communications from collectors, notes from museums, etc., was painstakingly and meticulously abstracted and compiled. The result, in an effort that spanned more than 20 years and countless thousands of hours, assimilated Eckel's original data, and incorporated the nearly 40 years of data that had appeared since his original work. Completed in 1997, the "update," as it came to be known, was published as *Minerals of Colorado*, by Edwin B. Eckel, updated and revised by Robert R. Cobban, Donley S. Collins, Eugene E. Foord (now deceased), Daniel E. Kile, Peter J. Modreski, and Jack A. Murphy.

This *magnum opus* constitutes a definitive resource for those investigating the mineralogy of Colorado. It includes 774 species (encompassing new 329 additions in the nearly 40 years since publication of the original work by Eckel) in a 665-page large-format book that includes color and black and white photos, locality maps indexed to descriptions of the major mining and mineral localities in the State, and more than 2,500 references encompassing the most significant literature on Colorado minerals.

It is unfortunate that Ed did not live to see the result of his work; I hope he would have been pleased. –Daniel Kile

Tickled
Pink
Rhodochrosites
on Quartz from the
San Juan Triangle
Kristalle est 1971
Wayne & Dona Leicht
Laguna Beach, CA, USA ~ +1.949.494.5155
info@kristalle.com ~ www.kristalle.com
Crystal Classics
Ian Bruce ~ East Coker, Somerset, England ~ +44 (0) 1935 862673
ian@crystalclassics.co.uk ~ www.crystalclassics.co.uk

Hardy, R.M. 1940. *Sunshine Mining Company Annual Report for 1939.*

Hardy, R.M. 1945. *Sunshine Mining Company Annual Report for 1944.*

Hillebrand, J.R. 1957. "The Idarado Mine." In *New Mexico Geological Society Guidebook Number 8: Southwestern San Juan Mountains*, edited by F. Kottlowski, 176-188.

Holmes, Richard W., and Marrianna B. Kennedy. 1983. *Mines and Minerals of the Great American Rift (Colorado-New Mexico)*. New York: Van Nostrand Reinhold.

Hoover, A. 1973. "Shrinkage Stoping." In *SME Mining Engineering Handbook* vol. 1, 12:135-140, edited by I.A. Given. New York: Society of Mining Engineers.

Hutchinson, R.M. 1988. "Structure and Ore Deposits of the Camp Bird Mine, Ouray, Colorado." In *Epithermal Base-Metal Systems, San Juan Mountains, Colorado.* Volume 3 of a series of guidebooks prepared for the Society of Economic Geologists field conference, October 26-29.

Ingersoll, Chester A. 1910. *Mining Report on the Grizzly Bear Mine and Other Continuous Properties*. Unpublished.

Jones, William R. 2007. History of Mining and Milling Practices and Production in San Juan County, Colorado, 1871-1991. In "Integrated Investigations of the Environmental Effects of Historical Mining in the Animas River Watershed, San Juan County, Colorado," edited by S.E. Church, Paul von Guerard and S.E. Singer, 43-85. U.S. *Geological Survey Professional Paper 1651* plus CD-ROM.

Kelley, V.C. 1946. "Geology, Ore Deposits, and Mines of the Mineral Point, Poughkeepsie, and Upper Uncompaghre Districts, Ouray, San Juan, Hinsdale Counties." *Colorado Scientific Society Proceedings*, 14:7.

Kile, Daniel E., Pete J. Modreski, and Dianne L. Kile. 1991. "Colorado Quartz: Occurrence and Discovery." *Rocks & Minerals*, 66:374-406

King, W.H. and P.T. Allsman. 1950. "Reconnaissance of Metal Mining in the San Juan Region." *U.S. Bureau of Mines Information Circular 7554*.

Larsen, E.S., Jr., and W. Cross. 1956. Geology and Petrology of the San Juan Region, Southwestern Colorado. *U.S. Geological Survey Professional Paper 258*.

Lipman, Peter W. and William C. McIntosh. 2008. "Eruptive and Noneruptive Calderas, Northeastern San Juan Mountains, Colorado: Where Did the Ignimbrites Come from?" *GSA Bulletin*, 120:7-8; p. 771-795.

Marshall, John, and Zeke Zanoni. 1996. *Mining the Hard Rock in the Silverton San Juans: A Sense of Place, A Sense of Time.* Silverton: Simpler Way Book Company.

Mayor, John N. 1978. "Geologic Summary of the Idarado Mine, Ouray County, Colorado." In Guidebook on Fossil Fuels and Metals, Eastern Utah and Western-Southwestern-Central Colorado, edited by D.R. Shawe, 131-140. *Colorado School of Mines Professional Contributions 9*.

Muntyan, Barbara L., and John R. 1988. "Recent Collecting Activity in Ouray County, Colorado" *The Mineralogical Record*, 19:89-102.

Muntyan, Barbara L., and John R. 1990. "The World of Rhodochrosite." *Denver Fall Show Guide*, 3:4-7.

Murphy, Jack A. 1979a. "The San Juan Mountains of Colorado." *The Mineralogical Record*, 10:349-361.

Murphy, Jack A. 1979b. "Mineral Collecting at the Sunnyside and Idarado Mines." *The Mineralogical Record*, 10:385-392.

Murphy, Jack A. 2003. "Preserving Minerals and Mining History at the Eagle Mine, Gilman District, Colorado." *Rocks & Minerals*, 78:324-328.

Murphy, Jack A., and Peter J. Modreski. 2002. "A Tour of Colorado Gemstone Localities." *Rocks & Minerals*, 77:218-238.

Nash, J.T. 1975. "Fluid Inclusion Studies of Vein, Pipe, and Replacement Deposits, Northwestern San Juan Mountains, Colorado." *Economic Geology*, 70:1448-1462.

Pearl, Richard M. 1972. *Colorado Gem Trails and Mineral Guide.* Athens: Swallow Press.

Purington, C.W. 1898. "Preliminary Report on the Mining Industries of the Telluride Quadrangle, Colorado." *U.S. Geological Survey 18th Annual Report*, part 3:751-848.

Raines, E. 1997. "Colorado Gold, Part II: The Discovery, Mining History, Geology, and Specimen Mineralogy of Selected Occurrences in Central Colorado and the San Juans." *Rocks & Minerals*, 72:304-320.

Raines, E. 2000a. "The Mineralogy, Geology, and Mining History of the Telluride District, San Miguel County, Colorado, Part I." *Rocks & Minerals*, 75:318-342.

Raines, E. 2000b. "The Mineralogy, Geology, and Mining History of the Telluride District, San Miguel County, Colorado, Part II." *Rocks & Minerals*, 75:378-401.

Ramsey, R.H. 1973. *Men and Mines of Newmont: A Fifty-Year History*. New York: Octagon Books.

Ransome, F.L. 1901. "A Report on the Economic Geology of the Silverton Quadrangle, Colorado." *U.S. Geological Survey Bulletin 182*.

Rich, Beverly. 2010. "Otto Mears, Pathfinder of the San Juans." *The Silverton Magazine*. http://www.silvertonmagazine.com/history/Otto_Mears.html.

Rickard, T.A. 1907. *Across the San Juan Mountains*. San Francisco: Dewey Publishing.

Rockwell, Wilson. 1998. *The Utes: A Forgotten People*. Lake Ciy: Western Reflections.

Rosemeyer, Tom. 1988. "Sunnyside Mine, Eureka Mining District." *Rocks & Minerals*, 63:366-384.

Rosemeyer, Tom. 1989. "Microminerals of the San Juan Mountains, Southwestern Colorado." *Rocks & Minerals*, 64:140-145.

Rosemeyer, Tom. 1990. "The History of the Camp Bird Mine, Ouray, Colorado." *Rocks & Minerals*, 65:114-49.

Rosemeyer, Tom. 1995. "Update on the Mineralogy of the San Juan Mountains, Southwestern Colorado." *Rocks & Minerals*, 70:180-187.

Rosemeyer, Tom. Unpublished. *History of the Longfellow Mine, Red Mountain Mining District, San Juan County, Colorado.*

Smith, A.E., Jr., and Tom Rosemeyer. 1986. "Great Pockets: The National Belle Mine." *The Mineralogical Record*, 17:229-236.

Smith, A.E. and Richard A. Kosnar. 1983. Across the San Juan Mountains by T.A. Rickard, part 2. *The Mineralogical Record*, 14:243-249.

Smith, Duane A. 2000. *Song of the Hammer and Drill: The Colorado San Juans 1860-1914.* Boulder: University Press of Colorado.

Smith, Duane A. 2009. *The Trail of Gold and Silver: Mining in Colorado, 1859-2009.* Boulder: University Press of Colorado.

Smith, P. David. 1990. *Ouray Chief of the Utes*. Ridgway: Wayfinder Press.

Smith, P. David. 2004. *Exploring the Historic San Juan Triangle*. Ridgway: Wayfinder Press.

Steven, Thomas. A., et al. 1974. *U.S. Geological Survey Miscellaneous Geologic Investigations Map I-764*, scale 1:250,000.

Stevenson, John J. 1875. "Report on the Geology of a Portion of Colorado Examined in 1873." In *Geographical and Geological Explorations And Surveys West of the One Hundredth Meridian, in Charge of First Lieutenant George M. Wheeler*, IV:303-501. Washington: Government Printing Office.

Strong, William K. 1988. *The Remarkable Railroad Passes of Otto Mears*. Silverton: San Juan County Book Co.

Trennert, Robert A. 2001. *Riding the High Wire: Aerial Tramways in the West.* Boulder: University Press of Colorado.

Vanderwilt, J.W. 1947. *Mineral Resources of Colorado*. Denver: State of Colorado Mineral Resources Board.

Vhay, J.S. 1962. "Geology and Mineral Deposits of the Area South of Telluride, Colorado." *U.S. Geological Survey Bulletin 1112-G*, 209-310.

Wolle, Muriel. 1977. *Timberline Tailings: Tales of Colorado's Ghost Towns and Mining Camps*. Chicago: Swallow Press.

Yager, Douglas B., and Dana J. Bove. 2002. "Generalized Geologic Map of Part of the Upper Animas River Watershed and Vicinity, Silverton, Colorado," part of the *U.S. Geological Survey Miscellaneous Field Studies Map* series. Denver: U.S. Geological Survey.

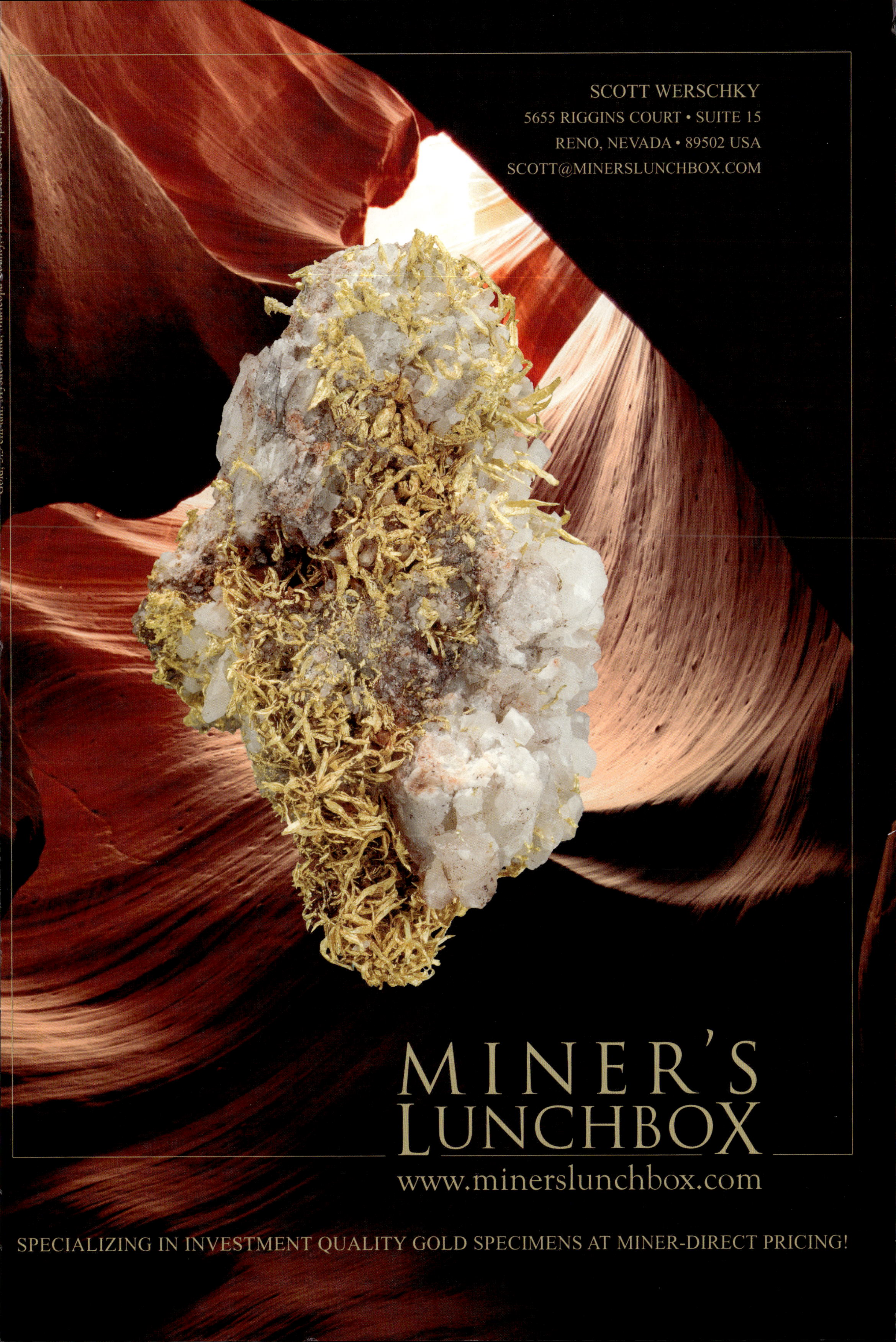
SCOTT WERSCHKY
5655 RIGGINS COURT • SUITE 15
RENO, NEVADA • 89502 USA
SCOTT@MINERSLUNCHBOX.COM
MINER'S
LUNCHBOX
www.minerslunchbox.com
SPECIALIZING IN INVESTMENT QUALITY GOLD SPECIMENS AT MINER-DIRECT PRICING!

Jesse La Plante photo

Worthy of Attentio

Colorado Rocks
COLORADO TOPAZ – Jesse La Plante, photo
DAVE BUNK MINERALS
1441 WEST 46TH AVENUE • UNIT 8 • DENVER COLORADO 80211 • 1.303.477.6897
WWW.DAVEBUNKMINERALS.COM